T0293804

ROUTLEDGE LIBRARY EDITIONS:
URBAN PLANNING

Volume 16

URBAN HOSPITAL LOCATION

URBAN HOSPITAL LOCATION

LESLIE MAYHEW

Routledge
Taylor & Francis Group

LONDON AND NEW YORK

First published in 1986 by George Allen & Unwin (Publishers) Ltd

This edition first published in 2018
by Routledge
2 Park Square, Milton Park, Abingdon, Oxon OX14 4RN

and by Routledge
711 Third Avenue, New York, NY 10017

Routledge is an imprint of the Taylor & Francis Group, an informa business

British Library Cataloguing in Publication Data
A catalogue record for this book is available from the British Library

ISBN: 978-1-138-49611-8 (Set)
ISBN: 978-1-351-02214-9 (Set) (ebk)
ISBN: 978-1-138-49005-5 (Volume 16) (hbk)
ISBN: 978-1-351-03606-1 (Volume 16) (ebk)

Publisher's Note
The publisher has gone to great lengths to ensure the quality of this reprint but points out that some imperfections in the original copies may be apparent.

Disclaimer
The publisher has made every effort to trace copyright holders and would welcome correspondence from those they have been unable to trace.

URBAN HOSPITAL LOCATION

Leslie Mayhew

Operational Research Service,
Department of Health and Social Security, London

London
GEORGE ALLEN & UNWIN

Boston   Sydney

George Allen & Unwin (Publishers) Ltd,
40 Museum Street, London WC1A 1LU, UK

George Allen & Unwin (Publishers) Ltd,
Park Lane, Hemel Hempstead, Herts HP2 4TE, UK

Allen & Unwin Inc.,
8 Winchester Place, Winchester, Mass. 01890, USA

George Allen & Unwin Australia Pty Ltd,
8 Napier Street, North Sydney, NSW 2060, Australia

First published in 1986

ISSN 0261-0485

British Library Cataloguing in Publication Data

Mayhew, L. D.
    Urban hospital location.—The London research
series in geography, ISSN 0261–0485; no. 4)
1. Hospitals—Great Britain—Location
I. Title   II. Series
362.1′1   RA967.7
ISBN 0-04-362054-X

Library of Congress Cataloging in Publication Data

Mayhew, L. D. (Leslie, D.)
    Urban hospital location.
(The London research series in geography, ISSN 0261–
0485; 4)
Revision of thesis (Ph.D)—University of London,
1979.
Bibliography: p.
Includes index.
1. Hospitals—Location—Mathematical models.
2. Hospitals—Planning—Mathematical models.
3. Hospitals—England—London—Location—History.
I. Title.   II. Series.   [DNLM: 1. Demography.
2. Health Services Accessibility.   3. Hospital
Planning.   4. Hospitals, Public.   WX 27.1 M469u]
RA967.7.M39   1985   338.6′042   85-15820
ISBN 0-04-362054-X (alk. paper)

Set in 10 on 12 point Bembo by Spire Print Services, Ltd, Salisbury, Wilts
and printed in Great Britain by Biddles Ltd, Guildford, Surrey

Preface

Most of the work and ideas presented in this book are based on a much longer doctoral thesis submitted to the University of London in 1979. The detailed research was carried out between 1975 and 1978 in the Department of Geography, at Birkbeck College, and was supported by a grant from the Social Science Research Council. The idea of converting the thesis into a book came from my supervisor, Dr P. Lewis, who had been strongly supportive throughout the duration of the research. Initially, it was hoped that the conversion would take only a matter of months. In the event the project has taken longer than the original research. There are one or two important reasons for the delay and I ought to say a few words on each because, to some extent, they also account for several refinements and important additions to the original thesis.

The first and, without any doubt, most important reason has been the unexpected interest in the subject of hospital location and this, not surprisingly, has tended to attract my attention away from the theoretical questions addressed in the thesis to the problems of applying and implementing some of the results. With the costs of health care soaring, the problem of providing a cost-effective and equitable pattern of health services is now an issue of considerable interest and importance to many countries. One dimension of this problem is the geographical question of exactly where hospitals should be located in relation to the populations they serve; another is what services they should provide and in what quantities. In cities, which are the focus of attention in the book, the difficulties of finding satisfactory answers to these questions are magnified by the shifting spatial distribution and demographic structure of the population. My first taste of experience with more practical issues was in 1979 when I joined the Operational Research Service of the Department of Health and Social Security. This was a period during which the London Health Planning Consortium was actively re-examining the strategic development of hospital services in the city. My main task then was to develop a spatial model to predict the flow of patients to hospitals resulting from changes in the supply of and demand for acute inpatient services. At the International Institute for Applied Systems Analysis (IIASA) in Laxenburg, Austria, between 1980 and 1982, I continued the development and application of this and related models as part of the health-care systems project. This was an especially creative and busy period, during which I was fortunate to be able to work with Georgio Leonardi, who at the time was leading another project concerned with the problems of public facility location. His expertise in mathematical modelling added another facet to my knowledge of location theory, which is reflected in two joint

professional publications. Since then, I have gained further work experience at
the Istituto Ricerche e Sociali in Turin, Italy, the DHSS again, and in 1984
with the Department of Mathematics at the University of Newcastle,
Australia.

This postdoctoral experience can be seen as advantageous in one sense,
because of the first-hand knowledge I was able to gain of health care systems in
different countries. However, in another sense, it can also be viewed as
disadvantageous, because of the problem of distilling some of what I had learnt
for inclusion in the text of the book, while remaining more or less faithful to
the ideas contained in the original thesis. I soon discovered that there is a world
of difference between having several apparently good ideas and then later
applying them in practice. And now, as a health systems analyst, I am very
conscious of the fact that ideas are just the first step in a very long development
process which ends in implementation. Of course, it was probably
unavoidable that I would need to include certain parts of this subsequent work
anyway because, in science, it is always necessary to refine a theory in the light
of experience. However, I have refrained from entering the realms of health
care systems modelling, which has been one of my major interests since 1980,
in too much detail. I hope that this work, of which only selected glimpses are
given in Chapter 4, will appear as a separate volume in the not-too-distant
future.

Finally I should say, of course, that this is a book which was written with
geographers primarily in mind. It was inspired, in large measure, by the work
of the German geographer, Christaller, who was responsible for the
development of Central Place Theory through his work published in 1933, but
not translated into English until 1960, entitled *Central places in southern
Germany*. His immense contribution to the field of location theory has long been
recognised within geography, planning, and to a lesser extent economics, and I
am well aware of the debt this book owes to him. However, I do hope that
others, including health planners, administrators, and operations researchers,
who probably never have heard of Christaller, will find the approach I have
adopted to be of some interest. Last but not least, there are two
acknowledgements I would like to make. They are to Peter Lewis, my PhD
supervisor, who has patiently awaited various earlier drafts of chapters, some
written in hotel rooms and on trains as I travelled around Europe during my
IIASA days; and to Geoffrey Hyman, with whom I worked for a short spell at
IIASA and more recently in London, for providing some valuable
comments on Chapter 3 in the last stages of the book's completion (some of
the material contained therein draws on a joint professional publication from
1982).

LESLIE MAYHEW

Contents

List of tables

To Karin

1 Exploring the problem and establishing a framework

1.1 Introduction

The extent of the contribution of health care services to the well-being of the human population is not known with certainty, but it is certain that we would be far worse off without it. Nevertheless, health care – the collective service rendered by physicians and other qualified personnel – is becoming an increasingly expensive activity with between 4 and 10 per cent of the gross national products in many countries now allocated to it. At a time when other sectors of national economies are making similar demands on public money, the question today is no longer whether we should be spending more on health care, but whether we are making effective use of what is already being spent. Aspects of this question fuel the entire debate over health care provision and range from the cost-effectiveness of current medical research, and the spiralling costs of modern medical technology, to problems of overmanning in parts of the administrative structure.

The geography of health care systems is rarely given much importance and yet, as this book tries to show, the availability of health services to the population in terms of hospitals, clinics and other facilities is strongly influenced by their location. Indeed, with a better geographical organisation of health care resources, many other problems associated with health services might resolve themselves. Nevertheless, there are fundamental obstacles: hospitals, which handle most of the demand for medical care, involve large financial investment and must be placed well in advance. Their locations and the shape of the buildings, once settled, are virtually fixed for their effective life as hospitals, regardless of population shifts and changes in other factors. It is rarely possible to estimate the effects of these influences without some prior conception of the underlying forces that motivate a health care system.

1.2 The urban dimension

Nowhere is the problem of organising hospital services more apparent than in large cities where increasingly more of the world's population lives (Gross 1972, LHPC 1979). (Approximately 1.8 billion people, 42 per cent of the world's population, live in urban areas today. According to a United Nation's estimate, 3.1 billion people will be living in urban areas by the year 2000). In 1971, for example, the London region contained over 500 medical institutions of all types and sizes delivering health care services. It seems reasonable to ask in what sense this is too much or too little; whether the existing locations reflect the present distribution of population and demand; and if they do not, what do they reflect instead and should they be changed; and finally, if change is necessary, what are the sensible methods of accomplishing it? These questions are fundamental to the provision of health services, but underlying them are very difficult problems related to the general functioning of all cities. Principal among these are the strong inter-dependencies between the three main economic sectors of the city: the sources of employment, the households, and the transport system linking sectors one and two. Phenomena such as urban sprawl, inner-city decline, housing decay, poverty and deprivation are some outcomes of the interactions between the three sectors that, in one way or another, affect health service provision.

It is useful to make a broad distinction at this stage between societies that are currently urbanising and those which are de-urbanising. In the former, supply problems in health care systems often arise because of the needs generated by the vast influx of rural poor into urban areas. The opposite is increasingly true in the latter where the stock of urban health care facilities was determined by earlier population influxes so that the subsequent reversal in the flow of population to the surrounding areas has created a relative excess of resources in the older centres. As the population redistributes to settle at lower densities, a relative deficit of health care facilities appears at the expanding city perimeters. In the sense it is used here, the term 'de-urbanisation' does not necessarily imply a fall in the population of a city. Indeed, the city may continue to grow in area *and* in population. The main criterion is that *some* areas of the city experience significant losses of population. In other words, de-urbanisation implies shifts in the concentration of people.

Some indication of the magnitude of the emergent problem of health care provision in cities is provided by the United Nations (1980). It notes that by the year 2000 there will be 439 cities with populations greater than 1 million, as compared with 185 cities in 1975; whereas the number of cities with more than 4 million people will increase to 86 (30 in 1975) and a further 25 cities will contain greater than 10 million. It is also noteworthy that by the year 2000 many of the largest cities will be in less developed countries where the problems of providing adequate health care are already immense. As a background to the rates of change that are taking place and to the scale of the emergent problem, Table 1.1 gives details of past and projected populations of some of the world's largest cities and their surrounding regions. The figures are indicative not only of the dynamic nature of urban areas and their populations, but also of the potential problems of catering for the health care needs of a growing as well as a shifting population.

Clearly, the scale and enormous cost of the resources required to provide an adequate level of service (currently about £4 billion a year in London and the surrounding region) demands careful planning. However, apart from the population implications, there are numerous other factors that impact on the question of health care provision. For example, the processes of urbanisation are linked to economic growth and changes in standards of living. These tend to affect the impressions of what people expect from a health care system in terms of the type and volume of services provided. Improvements in medical techniques, better access to information, increased mobility and hence spatial choice all create a heightened sense of consumer awareness, and these help to intensify the pressures and demands placed on health authorities and agencies.

Eventually, the effects of these separate influences – urbanisation and consumer awareness – are reflected in new health care facilities and changes in the spectrum of services offered at different locations. In the interim, the rapidity of change, and the complexity of the processes of change, raise numerous economic, social and political issues. Hospitals are very expensive to build, maintain and equip, and take a long time to plan and construct. As the

Table 1.1 Past and projected populations of some of the world's largest cities (in 10^6 persons).

	1950	1975	1980	2000
New York	12.3	19.8	21.8	22.8
Tokyo	6.7	17.7	23.4	24.2
Paris	5.5	9.2	10.9	11.3
Moscow	4.8	7.4	8.5	9.1
Calcutta	4.4	7.8	11.9	16.7
Los Angeles	4.0	10.8	13.3	14.2
Mexico City	3.0	11.9	22.9	31.0
Greater Bombay	2.9	7.0	12.0	17.1

Source: United Nations (1980).

system slowly adjusts to keep abreast of change there are bound to be some areas that are temporarily receiving less than their fair share of provision and others that are relatively over-provided. In some cases, the closure of older hospitals in declining areas of the city may have adverse effects in terms of access and convenience on those groups, usually the poor and old, that are left behind. Such effects as these often result in acrimonious debates as services are withdrawn, but they are rarely viewed in the context of a fully integrated city-wide hospital system. Such a scheme would allow more scope for service specialisation in different locations, a more efficient or equitable geographical coverage of services, and an interlocking service structure. The problem to solve, as far as this book is concerned, is what such a system might look like in idealised form and to what extent it is being reflected in the hospital system of today.

1.3   Scope and objective of the study

In more concise terms, the basic objective is therefore to develop a theoretical framework for exploring patterns of hospital provision from the standpoint of location. The aim is to consider the circumstances in which the location issue is important and to determine the broad consequences arising from good or bad spatial patterns of provision. The framework is designed to cope with a fairly wide class of cities, so that relatively little attention is paid either to specific details of particular administrative organisations in different countries, judgements concerning the relative merits of particular hospitals and the detailed specification of the services they provide, or technical issues of a medical nature. The idea is simply to make a detailed evaluation of the dominant considerations underlying hospital location, taking into account the shape and area of typical cities, the size, distribution and structure of their populations, and the relative ease of access between different areas and hospitals. Because it is an idealised framework, the simple models that emerge are of limited direct value for solving specific questions arising in particular cases. They are, however, of general value for considering the relative merits of current and future patterns of provision, particularly on a longer term, strategic basis.

To obtain insights into the usefulness of the framework in terms of its explanatory power, past and contemporary patterns of hospital provision are compared and analysed using the city of London as an example. Although this city is now de-urbanising, its development during the 19th century exhibits numerous parallels with contemporary urbanisation patterns in many developed and less developed countries. Indeed it is possible to observe almost an entire evolutionary cycle – from the take-off of the hospital movement more than 150 years ago, through a phase of rapid growth in the late 19th and early 20th centuries, to the more recent demise, decay and eventual closure of many old hospitals. Thus, it is hoped that by considering the hospital system of this city in detail, some important general principles can be developed.

Because health care systems are highly complex and because, even within the limitations of the study outline described, there is an extremely wide range of important issues, it is also useful to focus at an early stage on the sorts of questions to be considered in later chapters and to indicate why these questions are important. For instance, one of the most important problems involved in locating hospitals is how to determine the relative advantages and disadvantages of different locational patterns. Clearly, a given number, size and geographical arrangement of hospitals has different implications for patients and providers, particularly in terms of the ease with which services can be accessed and the costs of providing them. Thus, a small number of large hospitals might be more economic for the provider, but it will create hardship for some patients because of the increased travel entailed. It is necessary, therefore, to examine the geographical consequences and economic bases of different locational arrangements to determine whether some are more sensible than others in meeting health needs.

Directly related to this problem is the question of whether hospitals should provide the same services as each other or whether there is scope for specialisation. If there is scope – because, say, the city is sufficiently large and there are adequate means of transportation between different areas – the issue then is to determine where the more specialized services should be located in relation to other services. Also, there are hospitals which have completely different types of locational behaviour because the services they provide do not need to be very close to the populations they serve. It is useful to identify these hospitals and to suggest what types of locations instead are more suited to their functions. In this way, methods may be developed to ensure the right locations are reserved for the most appropriate hospital services. As a final instance of why the locational perspective is worthy of attention, one needs to consider again the impact of urbanisation on the spatial distributions of the population and hospital services. When supply and demand are not in balance, as is often the case in cities, it is important to know which types of hospital risk closure and to what extent the level of risk depends on the size, age and location of a facility. This will provide a basis for determining methods of adaptation to other functions as alternatives to closure.

These are examples then of problems requiring further investigation which provide some of the motivations for examining the question of hospital location. However, to determine the relative merits of different proposals, the locational impact of different health care issues, and what these mean in the case of a typical city, it is necessary to be more precise about the general factors governing the functioning of a health care system as a whole. This is because they differ in many respects from other activities and services which might otherwise be considered to have a similar locational behaviour. From these factors must be separated those which are most likely to motivate a health care system through space and time and which can provide a suitable basis on which to build the appropriate analytical framework.

1.4 Locational background to hospital provision

Whereas hospital provision in many countries today is extensively planned, the foundations of the earliest health care systems were left essentially to chance, so that the detailed histories of individual systems tend to be very complex. Typically, they are characterised by an increasing number of regulatory controls, introduced by governments, professional societies and other bodies to increase resources, improve efficiency or prevent unsound medical practice. Parallel with these developments, there have also been considerable changes in the threats to health, such as the decline in incidence of infectious diseases and the rise in incidence of chronic diseases, substantial changes in medical technology that have revolutionised treatment practices, and economic changes in the way services are financed and managed (Eckstein 1958, Singer and Underwood 1962, Arrow 1963, Abel-Smith 1964, Reder 1965, Weisbrod 1968, Newhouse 1970, Culyer 1971). These developments have interacted with one another and with the urban environment in such a complex fashion that it would be very difficult to take them all into consideration. To make any progress, therefore, it is necessary to isolate from the detail a thread of continuity that tends to motivate behaviour of a health care system in space and time.

Perhaps the only obvious and consistent thread is that provided by the moral and ethical foundations of the health care system related to the care and treatment of the sick on the basis of 'perceived' need because this, traditionally, is the basis of health care itself. Certainly it is not overstating the argument to suggest that participants in health care systems have, on the whole, been motivated in their work and in their allocation of time and medical resources by this fundamental driving force. Perceived need, however, is not an objective measure that determines precisely how many resources are required or how they should be distributed. In addition to medical criteria, it must be understood in the context of the individual and the social, technical, political and economic environment existing at a particular time. It is of interest in this respect to consider briefly the stated aims of the National Health Service in England and Wales, since it is an embodiment of this linking theme and is also one of the few occasions that the aims of a health care system have been officially acknowledged in this way. It is '. . . to ensure that every man and woman and child can rely on getting all the advice and treatment and care they need in matters of personal health . . . [and] . . . that their getting these should not depend on whether they can pay for them' (Feldstein 1963, p. 22, quoting para 1 HMSO 1944 p. 5). There are, of course, complex issues concerning whether and how much and by what means health care should be paid for, but the expressed aim of having universal access is one now accepted by almost all countries. Nevertheless, it must be strongly emphasised that there are at least two serious problems associated with the above ideals that, in general, prevent their realisation.

The first of these is the assumption that all the health needs of a population can be catered for. On the basis of recent experience, this has proved to be wholly unrealistic. Despite improvements in the standards of living and health indicators in general, the demand for and cost of health care has continued to rise at an alarming rate, not only in England and Wales where these ideals were expressed, but also in many other countries with altogether different types of health systems. The result has been that, rather than cost of health care becoming cheaper through improvements in diet, hygiene and preventative measures, as was originally supposed, the opposite has actually occurred. The second fundamental problem is implicit rather than explicit in the statement of objectives. It is that, as long as patients pay in time, money, discomfort and other costs for access to hospital services, the fact that the services are nominally free or are of no direct cost to the patient does not eliminate the possible hardships caused by an unfair distribution of resources, poor access to facilities in some areas, or long waiting times for treatment.

Indeed, there is mounting evidence to suggest that even at relatively low geographical scales, such as in small areas within cities, disparities in provision can have a significant effect on demand (e.g. LHPC 1979). It is convenient for current purposes to consider these two problems as relating to two essentially different issues. The first, for example, would be one much more concerned with the functioning of an economy as a whole, especially in terms of how much total expenditure on health care can be afforded. The second, by contrast, is concerned more with the mechanisms with which resources are actually allocated and their effectiveness evaluated. As the cost of health care has grown, governments and health authorities have begun to focus increasingly on the distributional issues and less on the issue of fulfilling needs. The reason for this is to ensure better value for money and a fairer apportionment of available resources (e.g. RAWP 1976). Indeed, the emphasis in some countries is now on satisfying the relative rather than the absolute health needs of the population in recognition of the fact that the demand for health care may actually be insatiable. It is these aspects of the distributional problem that are concerned primarily with the inequalities and inefficiencies arising out of a poor location of resources that are of principal interest here.

Before continuing, it is important to reflect that the above objectives of a health care system are not the only ones. There are also many subsidiary objectives which contribute to the fulfilment of the main objectives and which need to be mentioned in passing. For instance, health care systems also have responsibility for training doctors, nurses and other staff, medical research, the developing and testing of new techniques and equipment, and so forth. These activities divert a lot of resources and, because they are concerned more with investment in future service patterns, they sometimes conflict with the current service needs of the population. For present purposes, these aspects can be regarded as a separate problem for the locational question, although they need to be borne in mind to prevent an oversimplified account of the activities and responsibilities of a health care system.

1.5 Central place theory

Existing locational theory provides a good basis for discussing the problem of locating hospitals in cities, but because of the special nature of the health care system it is deficient in several respects. This is because much of the subject is traditionally concerned with the problem of locating individual facilities, such as warehouses or factories, where profit or cost criteria form the basis for selecting a particular location. In the case of hospitals, the problem is much more concerned with the location of many facilities, how they function in relation to one another, how accessible they are to potential users and whether needs are being satisfied. Moreover, the available criteria for judging different locational patterns are not as precise as they are for the ordinary private firm, so that they must be seen within the context of the particular goals of the managing health authorities. This is because patients, the consumers of health care, are partly guided by their own judgement about whether to use particular facilities. Such judgements are determined by many factors, one of the most important of which is accessibility.

Largely because it deals with the spatial provision of services, the most relevant branch of location theory for current purposes is probably central place theory. This was developed originally by Christaller (1933, 1960) to explain the geographical organisation of metropolitan areas, cities, towns, villages and smaller settlements and the goods and services they provide to surrounding populations. However, because of the generality of the principles underlying the theory, further research was able to extend its range of applications to other geographical scales, including the spatial organisation of retail services within cities. This is at a scale closer to the proposed level of the present enquiry and so it is of interest to ask to what extent a similar approach can be applied to hospital services instead of retail services.

Common to most versions and applications of central place theory is a partition of a region into a regular hexagonal hierarchy of market areas. The exact form and reasons for this hexagonal hierarchy will become apparent from the many examples given later, but essentially the idea is as follows. Consumers have needs and preferences that are satisfied at different intervals depending on the good or service in question. Regularly demanded, low-order goods and services are obtained from the local stockist or supplier but for less frequently needed and more expensive high-order goods and services, a special trip is necessary to the next or subsequent layers in the hierarchy (e.g. from a village to a town or city). It is acknowledged that such hexagonal patterns are particularly efficient ways of delivering services to a dispersed population, because they are broadly consistent with economic behaviour and equity criteria. Thus, in the long run, competition eliminates duplicate or poorly located service facilities, and the residual traders can maximise their profits. At the same time, the population itself has an even access to goods and services throughout the region because, on the basis of the geometry of the system, no one is more than a fixed distance from the given service. In spite of this evenness, however, it is important to stress that, by concentrating services in specific locations a penalty results in the form of a reduced accessibility for some, and this influences the use of services accordingly. Plainly, hospital services are also hierarchical in nature: regional facilities, such as teaching hospitals, provide highly specialised services to a wide area, but also local services, whereas small hospital facilities, including clinics and health centres, provide only the latter. Why does this separation of functions come about in health care systems as well as in retail and related services and settlement systems?

The first step is to consider exactly what we mean by a 'health' service. Patients are treated in hospital on the basis of their clinical condition and any service they receive is therefore appropriate to that condition. They may, for example, use an operating theatre or X-ray facilities; at a later stage of their treatment they may require the services, say, of a physiotherapist. If they are also inpatients then they will also require so-called hotel services which are all those services pertaining to hospital accommodation such as bed provision and meals. The situation appears, therefore, not to be straightforward in the sense that the average patient consumes not just one service but a collection of

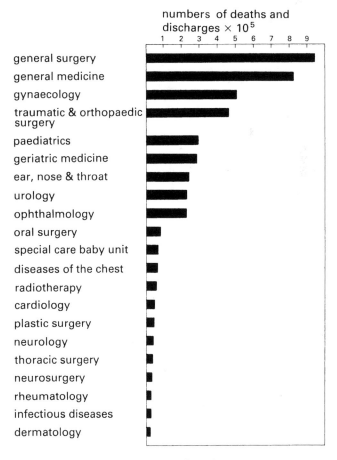

Figure 1.1 Estimated deaths and discharges from hospitals in England and Wales by department in 1978 (minor departments not shown).
Source: HMSO 1978, see Table 12, p. 220 and p. 442 for maternity cases.

services which together might be said to form a 'package' of services. The problem of defining a 'service' is greatly simplified if we observe that patients with broadly similar clinical conditions tend to be treated within the same department of a hospital. These departments are organised into specialty groups and, at a broad level of generalisation, the patients in them make similar demands on other, shared hospital services and facilities. This does not mean that a patient with a particular clinical condition is uniquely allocated to one particular specialty. Sometimes the definition of specialties can be very wide and, depending on the organisation of the hospital, this can lead to a substantial degree of overlap. Nevertheless, if we examine data on the pattern of hospital admission by specialty grouping, we can begin to observe the elements of a potentially hierarchical system, rather like the example of retail services. Figure 1.1, for example, shows a histogram of the estimated numbers of patients leaving hospital in England and Wales in 1978 in 22 of the most important (numerically speaking) clinical specialties. Some of these specialties, such as plastic surgery and dermatology, treat relatively few patients whereas others, such as gynaecology or general surgery, treat a great many. What most distinguish individual specialties are not only the types of patients, but also the special expertise of the medical staff, and the techniques and equipment they use. For minor specialties attracting relatively few patients this means that the costs of provision are such that they are often uneconomic to provide at the very local geographical level. Located at the intermediate or regional level, however, they are able to generate more patients per unit of resource deployed. Moreover, they can share many of their overhead costs such as ancillary services with other, larger specialty groups within the same hospital. This initial overview of the functioning of a multilevel hospital system, then, is the most obvious link with central place theory and it is a theme being increasingly employed in the health care literature (e.g. Schultz 1970, Shannon & Denver 1974, Dietrich 1977, Parr 1980).

1.6   Some particular deficiencies in the theory

There are several difficulties with the above arguments, but they are unrelated to the aspects we want to criticise and develop. However, they should also be stated in passing since it is important not to convey an oversimplified picture of what is occurring. Firstly, because there are qualitative differences among the same services provided in different locations reflecting resource availability, treatment practice, expertise and other factors (Feldstein 1965b), the services may not be strictly comparable. Secondly, which services are provided where and in what level is not likely to be constant, certainly over the time span we shall be considering. Just as other parts of the economy are affected by technological change and changes in relative resource costs, so are hospital services. As the total budget for health care increases or decreases, as new techniques are introduced, or as other factors change, so different services diffuse at varying rates through the hospital system (Abel-Smith 1964). However, when we look at hospital location in cities, and also at the underlying factors motivating the patterns of supply and demand, and then try to apply central place theory, the above considerations are very much overshadowed by four new ones, which are now discussed in turn. The first two of these would apply to almost any health care system, whether rural or urban, but the last two are specific to the types of locational environment found in many cities.

There are few market signals regulating the system
The first problem is that there are few market signals to allocate and reallocate resources in a health care system, or to regulate supply and demand. This is because patients pay little, if anything, directly for their treatment except in terms of waiting or travel time, opportunity costs (e.g. time lost at work) and possibly discomfort. The financial costs that arise are usually met by insurance companies by direct transfer to the provider of the service or by the state, if the system is nationalised, whereas in central place theory the level of service provision is determined solely by the profit criterion. If a service does not generate sufficient customers to cover its costs, it is withdrawn or taken up one or more levels in the hierarchy, but in a health care system this would not necessarily happen because the appropriate regulatory mechanisms are much less simple. This is because there are enormous difficulties in comparing the relative and marginal benefits of different services or treatment procedures in the sense that the return or output from a unit of investment, except in a very limited number of instances, is intrinsically unquantifiable (e.g. Weisbrod 1968, Illich 1976). Moreover, in clinical matters, the preferences of patients are influenced less by cost considerations (since they are unaffected by them) and more by medical opinion which, in turn, depends on the state of the medical art (Fuchs 1966). These and other factors make it hard to judge from the patients' and providers' viewpoints, the right levels of provision overall, and the balance between the allocation of medical resources to different services or sectors of the health care system.

The need for health care is ill defined
Because they are insured against medical costs, most patients incur only minimal costs at the points and times they use health care services and this affects their use of them. In predicting demand, measures based on health status ought, therefore, to be more reliable than, say, the level of a patient's income because, the argument goes, patients would base their usage of health services on strict medical needs. However, considerable uncertainty surrounds the usefulness of this and other measures aimed at identifying appropriate levels of resource provision because experience indicates that such measures tend not to be related strongly to *actual* demand. For example, the patterns of

usage that operate in a locality tend to be influenced by the local availability of health services as well as by need, so that, for the analyst or professional, it is never clear to what extent demand is a reflection of genuine needs. This is not particularly a problem if patients are simply exercising their rights as taxpayers or insured individuals, but it may be a problem if the resources deployed could be used more effectively elsewhere. In central place theory, in contrast to this complex behaviour, the demand for services is already known and the level of supply is assumed to adjust accordingly. This seems reasonable in the case of retail services for which the level of supply in a locality is more closely identified with the income levels of potential shoppers. In health services, by contrast, demand is considerably more elastic and dependent on many factors, not only health needs.

There are complex variations in the spatial distribution of demand in cities
There are significant differences in population density between the centres and peripheries of cities and this influences the way hospitals are located. As urbanisation or de-urbanisation proceeds, the spatial distribution of demand changes and the existing distribution of hospitals and the services they provide are disturbed in a variety of uncertain ways due to losses or gains in catchment populations. In central place theory, both the population and the service facilities need to be evenly distributed and locationally fairly constant over time to ensure an equality of access and service efficiency, and for the distinctive hexagonal market areas to evolve. Except in limited circumstances, the theory cannot deal with the structural and spatial changes in supply and demand that typically arise in realistic geographical cases. In a city, if a facility is forced to close due to falling population, then a 'hole' appears in the system leaving the surrounding locality unprovided. The growth of a central place hierarchy as services at centres crystallise is not accompanied by any precise mechanism for implementing this structure; similarly the decline in central services has no associated process of change. Central place theory is not a dynamic theory.

Accessibility patterns deviate from the theoretical patterns
Central place theory places a strict emphasis on the distance consumers are from supply as a factor controlling the demand for services. If we look at travel patterns within cities, it is evident, however, that other factors are affecting the apparent simplicity of the distance mechanism. The ease of travel, and thus the effect of the distance mechanism, is influenced by the pattern of routes, the level of congestion, the availability of public or private transportation and so forth (Wingo 1961). Typically, it might be supposed that an efficient transport system would create a much wider spatial choice in terms of available services, although it is questionable whether everybody benefits from this enlarged choice. Two broad categories of travel effects likely to influence the locations of hospital facilities can be recognised at this stage. They are the mode effect

due to the type of transportation used and the effects due to the physical and behavioural characteristics of individual patients. The first category refers to the nature and efficiency of the urban transportation system. The public transportation network, for example, has a strong radial bias and so tends to emphasise the attractiveness of locations in central areas of a city. Specialist hospitals in particular often locate at the centre because, from there, they can attract patients from a very wide area. Private transport, on the other hand, is considerably more flexible than public transport, offering a wider choice of routes and more possibilities for peripheral or 'sideways' travel. Central areas, however, are less attractive to motorists because of congestion and parking difficulties. The second set of effects based on the behavioural characteristics of the travellers are also of importance. Hospital patients consist partly of the elderly and infirm and, in certain cases, they experience hardship as a result of poor hospital access. Although some hospitals provide ambulatory services, these soon become very expensive if they are extended to large areas of the city. In summary then, the implications of both the modal and behavioural effects of travel in cities in relation to the distance mechanism in central place theory are both considerable and complex.

The above analysis emphasises some very serious drawbacks in central place theory, to the extent that they would appear to rule out its use entirely. However, it is important in thinking about the problem not to forget those features of the theory that make it so attractive to planners and geographical analysts and which are plainly an essential consideration in the structure of any health care system. They are the concepts of market area, hierarchy, service specialisation, geographical coverage, efficiency of provision and equity of access. The next section outlines, therefore, a set of modest changes to the theory that will be developed further in subsequent chapters. They are indicative of the sorts of modifications that must be carried through if the theory is to provide an adequate framework for the patterns that operate, or if it is to be used for planning purposes. In some respects it will be seen that the proposed changes do not go as far as they might, but whether or not they are sufficient for a general study of hospital location in cities must be judged from the case study later. What they try to provide is an extended framework for evaluating different locational patterns of facilities in a way that is arguably relevant in the context of cities today and which can be built on in further research.

1.7   Changing objectives and relaxing assumptions

It was noted above that the profit criterion used in central place theory is not a relevant consideration where providers of hospitals are concerned. Instead, health care systems are striving in some sense to meet the health care needs of the population. The first change that can be made to the theory, therefore, is to relax the profit assumption and consider the need objective instead. As discussion has indicated, however, the difficulty is to define this objective in relation to the processes of locating and allocating resources. Needs can be satisfied in a variety of ways and, unlike the profit criterion used in relation to retail services, there is no unique solution to the spatial problem of organising hospital facilities. The answer depends on the total resources available for allocation and the relative priority attached by health authorities to providing one pattern of services rather than another. Given this dilemma, therefore, one possible way forward is to define in very general terms a basic set of locational patterns which are justifiable on theoretical grounds as being, say, equitable or efficient, and then to examine the extent to which they satisfy health needs. The implications of steering the health system in one direction rather than another can then be more clearly understood. From the point of view of discussion, efficiency in this context is taken to mean the greatest possible economy in the use of resources required for the objective in question. Equity, by contrast, means allocating resources such that, to the extent desired, all people in a city have fair and equal access to available resources. From earlier discussion, and from examples given later, it is clear that central place theory is well qualified in both these respects in the sense of providing both an equitable and efficient solution, but only under restrictive conditions of population distribution and urban travel behaviour. When these conditions are not met, as is the case for most cities, the predicted patterns of spatial organisation yielded by different criteria diverge by differing amounts with different implications in terms of the services affected and the costs of provision incurred. Thus it is important to discuss the advantages and disadvantages of each criterion and to evaluate the consequences of each in relation to health care needs.

It was seen that a key feature of central place theory is the regular, hexagonal market area denoting the trading boundaries of a particular service. For hospital services, the use of the term market area is perhaps inappropriate because of its shopping connotations. Furthermore, we have seen that different organisational objectives are likely to yield different spatial patterns of hospital facilities, indicating that the hexagonal market area is simply one of many possible geographical outcomes. The second change we make to the theory, therefore, is to alter the name from market area to hospital district and to define the process of subdividing an urban area as districting, where it is assumed that each district in an urban area is served by one hospital. The idea then is that different districting criteria can be developed to reflect different organisational objectives of the health system and thereby enlarge the range of allocative options open to health service planners or administrators. A direct consequence of the second change is that, depending on the circumstances, certain other features of central place theory sometimes have to be sacrificed. Two features in particular are universal equality of access and the behavioural assumption that patients use the nearest hospital supplying a particular service. It is a fact, however, that both features depend on rather strong assumptions about accessibility and demand behaviour which, as was seen, need to be heavily qualified in any analysis of urban hospital provision. It seems sensible, therefore, not to be too insistent on their validity providing they do not endanger the essential theoretical structure and the implications of relaxing them are properly understood. This is important because it will allow us to explore and interpret a much broader set of spatial options.

The discussion of whether or not health needs are best satisfied by one pattern of provision rather than another cannot proceed independently of the behavioural basis of demand and how far patients travel to seek a particular hospital service. Earlier, we saw that demand is sensitive, among other things, to the spatial provision of services and it would seem sensible, therefore, to make allowance for this type of behaviour during later analysis. On the other hand, it would be unwise to rule out other forms of demand behaviour until they have been empirically tested. The third change we make to the theory, therefore, is to modify the way demand is presently handled. We do this by postulating two different forms of demand behaviour that are representative of the way demand is usually portrayed in other studies, and which offer scope to analyse a wide range of potential implications as far as hospital location is concerned. The forms of demand in question are termed fixed demand behaviour and elastic demand behaviour. The first form assumes, more or less as in the existing theory, that users express a demand for health services up to their needs, which are independent of the resources available, but who vary their use of a particular facility according to the difficulty of geographical access. The second form, by contrast, assumes that users are elastic in their demand for health services, which increases or decreases according to the local availability of treatment facilities. The key distinction then is that in the first case users are always able to satisfy their demands, if not in one location then in another, whereas in the second case a proportion of demand is always unmet. Although more details are given later, it is worth stressing that unmet demand does not necessarily imply untreated demand. Potential patients may receive equivalent treatment (though of a lower quality) from their family doctors, non-hospital organisations or through self-treatment. The elasticity component derives from the relative availability of treatment facilities (of whatever kind) and the severity of a patient's condition.

The fourth and final modification is concerned with the measurement of accessibility and how it can be made responsive to the special accessibility characteristics of cities. One approach is to measure accessibility in terms of travel time (instead of distance) which arguably is a more relevant indicator influencing the choice of hospital. Clearly, however, this step raises numerous difficulties because travel times are dependent on the time of day a journey is undertaken and the type of transport a patient uses. In addition, large cities contain many thousands of roads and, plainly, it would be unrealistic to establish information on the travel times by different modes of travel on every link. Instead, the approach adopted is to consider travel times and the associate choice of routes in a very generalised way. In particular it is assumed that, at a certain scale of geographic resolution, average travel speeds can be represented as varying continuously over the urban area. In particular, they can be described by a simple function of distance from the city centre known as a velocity field (Angel & Hyman 1976). The idea is that velocity fields can be used to characterise, at a general level, the relative ease of movement in each area of a city. Thus, in inner areas where the roads are narrower and congestion is higher the speed of traffic flow is portrayed as being lower than in outer areas where traffic is less. Of course, the use of velocity fields is more appropriate to some cities than others, and it is also suited to some modes of travel more than others. Plainly, travel by rail in the interstices of the rail network would be impossible so that, in this case, the continuity assumption is not met and another approach must be adopted. One important and related implication of the velocity field approach is that different spatial patterns of provision will be appropriate for different modes of travel. This is an important point because it further enlarges the range of spatial options available and hence the range of feasible districting patterns.

1.8   Hospitals and the problem of scale economies

Now that those aspects of central place theory that need to be changed or adapted have been considered, an associated question concerns the likely financial consequences of basing health facilities on one set of proposed locational patterns rather than another. It seems clear that these consequences will depend not only on the resource costs and the overall level of health care provision in an urban area, but they will also vary according to the economic scale effects of hospital size and the opportunities for making savings, for instance, by basing services in larger facilities. If it is supposed that the sizes of hospitals are scaled in proportion to the populations served, then clearly different districting criteria will generate differences in the total costs of provision, depending on the relative efficiency of individual hospitals in relation to the optimum size. The nature of the financial relationship between locational pattern and hospital size will therefore depend critically both on the total quantity of resources allocated and how they are distributed among different locations. For these reasons, it would be wrong to consider the problem simply as an exercise in determining a size of hospital that is the optimum for a particular city (e.g. Cohen 1967). The existence of districting criteria necessitating different allocations of health care resources, and the constraints imposed by accessibility costs, obliges health authorities to consider the costs of the system *as a whole*. This gives rise to a set of interesting theoretical problems which can be considered within the modified central place framework and which lead to some surprising conclusions regarding different possible patterns of provision. One further complicating factor, however, is that hospitals can economise on their costs through a reduction in the *quality* of the treatment they provide. This means that two apparently identical services in two locations may function along different economic lines. Such differences, however, may be regarded essentially as a data problem in the sense that it is important in any study to compare only like with like. In general and despite major advances in the development of statistical information, we are still some way from achieving that goal.

1.9 A simple geographical typology of cities

Before proceeding with a consideration of some of the more practical issues associated with the case study for evaluating the theoretical framework, it is necessary finally to be more precise about the geographical forms and sizes of the cities that will be considered and the mathematical notation needed to describe them. This is because not all cities will provide convenient locational environments in which to undertake similar applications of the proposed methods without qualification. For instance, an important consideration is that the cities concerned are quite large, so that there is scope for many hospitals, over 30 say, as well as for some degree of specialisation in terms of the services available at different locations. Although smaller cities can also be dealt with using the approach, they are less interesting because they do not present the same scale of problem. A further criterion is that, like London, which forms the basis for the case study, the cities will be central to a more or less circular region. They may be either compact or dispersed in areas, but usually they will be characterised by a centre to periphery decline in population density. The forms of city described are hence very similar to the conceptual models developed by eminent urban sociologists and economists, such as Park and Burgess (1925), Wingo (1961) and Alonso (1965). Besides London, other good examples of cities in this category include Paris, Rome, Moscow, Munich, Budapest and Vienna, all of which have the required circular form but which also vary in their degree of compactness and size of population. In cases where a city is located on a coast, by a lake or on an estuary, the approach is easily modified by assuming that a wedge or some other shaped section has been removed from the otherwise circular form. Examples of cities in this category include Chicago, Sydney and Marseilles. However, the cities not covered entirely by the framework at present are those that develop linearly along a highway, valley or coastal strip, such as Genoa in northern Italy or Brighton in England, or those which are more properly described as an amalgam or agglomeration of several cities with several major centres, such as Birmingham in England or the industrialised region of the Ruhr in West Germany. They represent a smaller class of problem, although some very interesting work remains to be done on them.

Because of the proposed concentration on the circular form, it is convenient to represent locations in a city by a system of polar coordinates (r, θ) defined over the urban area in which the location $(0, 0)$ represents the city centre and (r_{max}, θ) the city boundary, whereas it would be more appropriate to characterise linear cities by a Cartesian system of coordinates (x, y). The distribution of population for the circular form is defined by a density function for which the notation used is $D(r, \theta)$, where D is the density of population in persons per unit area at a point distance r and direction θ from the city centre. In the cities we shall mostly be considering, the population is presumed to be dispersed more or less symmetrically around the city centre, so that the decline in population density along any radial is broadly the same. This means that θ can often be omitted from the notation, in which case the density function can be abbreviated to $D(r)$. A similar argument can be used for describing accessibility characteristics of a city. For example, it is convenient to represent the velocity field, which characterises the ease of movement at different locations, by $V(r)$, where V is the average speed of travel at distance r from the city centre. The mathematical forms of both $D(r)$ and $V(r)$ and their validity as generalisations will depend on the physical attributes of the city, such as the concentration of housing and the pattern and efficiency of the transport networks. Nevertheless, neither $D(r)$ nor $V(r)$ is designed to capture fully the range and complexity of the factors influential in the provision and procurement of health care services. For a more complete evaluation it would be necessary to disaggregate both $D(r)$ and $V(r)$ in various ways depending on the aspect of the health care system under consideration. For example, the demand for health is very sensitive to the age and sex of the population, and if there are spatial variations in the demographic structure of the urban population, it would be more appropriate to re-specify $D(r)$ accordingly. Similarly, the ease of access in a city is, as noted earlier, affected by the mode of travel and the time of day at which a journey is undertaken. Such detailed considerations open up much wider areas of investigation and analysis and are only partially considered. Such detail must be left to further research, although there is no reason why, in dealing with these matters, the basic framework should be significantly altered.

1.10   A classification of hospitals

To conclude the introductory analysis, it is desirable to relate some of the ideas and concepts expressed in earlier sections to the practical considerations of devising a suitable case study. To begin with it is necessary to create an appropriate classification of hospitals, since not all will share the requirement in central place theory of customers travelling to a central point to obtain a service. Although there are other types of health care facilities besides hospitals, such as clinics and general practitioner surgeries, the emphasis in the case study is confined to the former. While other facilities could equally be considered as part of the chosen approach, their inclusion would add greatly to the exposition of the concepts and their application.

Perhaps the most important distinction to make in the classification scheme is the length of time a potential patient remains in a particular type of hospital. For example, in psychiatric hospitals the average length of patient stay may be so long and the hospital throughput so slow that there is arguably less necessity to locate these facilities in places where they are accessible to the population. Indeed, data from different countries show that the hospitals concerned are generally small in number, very large and located some distance from city centres. In their cases, it is implicit that long distances or high travel times are not an important constraint on their usage (although potential visitors to patients might be deterred from coming as often as they might). Accordingly, health authorities might emphasise in their selection of locations factors such as site costs or environmental aspects (such as solitude or fresh air) which are thought to be beneficial to patients. For these hospitals, therefore, the spatial requirements are currently regarded by the classification as being non-central. Although there are signs of change in this locational behaviour with the establishment of smaller, more accessible facilities, the practice, to date, has generally been for them to be located away from large concentrations of population. For the most part, other types of hospital have a contrary locational behaviour to the extent that they tend to be attracted to centres of population. For example, acute hospitals treating short-term patients and forming the largest group of hospitals conform rather well to the basic requirements of the central place approach. Typically, they are built in a wide range of sizes; and whereas the smaller hospitals provide a restricted range of low-order services of a general nature for use by the local community, the larger ones have additional, more specialised services attracting patients from farther away. Because acute hospitals are used and visited more frequently, their locations, unlike psychiatric hospitals, are much more sensitive to accessibility costs and this is why their locational requirements are defined as being central (as opposed to non-central).

In the study, acute hospitals are termed type 1 hospitals and because of their centrality properties, more emphasis is given to analysing their locational patterns. Nevertheless, it is still very important to compare, contrast and predict the locational behaviour of other types of hospitals because, together, they should form an integrated system of health care provision. Four other types of hospital are included in the classification scheme: they are specialised (type 2) hospitals, which have a high throughput, serve very large areas and are highly attracted to high concentrations of population; psychiatric (type 3) hospitals which, as noted, have low throughputs and are relatively remote from centres of population; long-stay (type 4) hospitals, which provide care for the old and chronically sick and whose locational requirements are intermediate between hospital types 1 and 3; and hospitals for infectious diseases (type 5) which, as will be seen, tend to have rather special locational

Table 1.2 A classification of hospitals.

Hospital type	Number of services	Order of services*	Throughput per bed	Locational requirements†
1. Acute	few to many	low to high	high	central
2. Specialist	few	high	high	more central
3. Mental	few	low	low to medium	non-central
4. Long-stay	few	low to medium	low	non-central
5. Infectious	few	low	sporadically high	to be determined empirically

*Degrees of specialisation in terms of services offered.
†Central – attracted to centres of population; non-central – not attracted to centres of population.

requirements. (The detailed definitions and sources of information used to develop the data base are described in an appendix elsewhere in the book.) The essential characteristics of the five types of hospital outlined above are summarised in qualitative terms in Table 1.2. The number of services, and the degree and range of specialisation associated with each, are broadly described in terms of low, medium or high; the locational requirements are defined in terms of centrality, which is simply a method of ranking different types of hospital in terms of their attractiveness to centres of population; and hospital throughput, the other important distinction, is also measured in terms of low, medium or high. It is noteworthy that Table 1.2 shows some uncertainty as regards the locational requirements of type 5 hospitals. This is because, historically, such hospitals had to be, on the one hand, accessible to the population and, on the other, reasonably remote from communities to avoid the dangers of spreading infection. Together, as will be seen, these factors result in a very distinctive set of locational patterns.

1.11   Hospital finance and administration: public or private

Apart from a classification of hospitals according to the degree of centrality, a second important dimension on which hospitals can be classified is according to whether they are publicly or privately financed and administered. In the modified version of central place theory presented here, this could lead to further contrary locational behaviour, particularly if there were a clash between profit- and service-based criteria. For example, under certain circumstances private hospitals might be more attracted to areas of high income potential, whereas public hospitals would be expected to base their locational decisions more on need-related criteria. In practice, as the case study shows, the problem is not quite so simple and whether or not variations exist will depend very much on the administrative and financial circumstances of the hospital systems in each city. In the case of London, hospitals in categories 3, 4 and 5 were, when they were built, firmly in the public sector. Acute and specialist hospitals, by contrast, were mostly independent, non-profit organisations with a variety of revenue sources, including public taxation, endowments and charitable grants. Thus, to attribute locational differences to administrative or financial criteria would be misleading since these differences would exist anyway due to the different types of services provided. A more fundamental question is whether similar categories of hospitals reach similar locational decisions if they functioned on the private or public domain. The answer to this question would depend, in the case of private hospitals, very much on the method of payment for treatment, the eligibility of different types of patient to use the facilities and the services provided.

A final, but interesting, complication to note is specific to London, and therefore not easily generalised to experience in other cities. This is the creation of the state-operated National Health Service (NHS) in 1948. Most hospitals, when given the choice of joining the new organisation, responded positively but a few elected to remain outside. As is shown in Chapter 5, however, there was some geographical selection involved among those declining to join, with certain areas of the city retaining a significant number of private hospitals. These considerations emphasise that it is necessary to be aware of historical factors that, from time to time, result in significant organisational changes. Among the references that were found to be useful in this respect were Eckstein (1958), Dainton (1961), Abel-Smith (1964), Pinker (1966), and Ayers (1971).

1.12 Conclusions

This introduction has provided the necessary background to the problem of locating hospitals in cities. A picture has emerged of an activity in which the benefits of health care provision are hard to measure, and in which the demand for services can be strongly influenced by non-financial considerations. It was also seen that, unlike other economic activities in which provision is taken care of by the profit criterion, this has been proved an imperfect mechanism in the case of health care systems. It was decided that the most appropriate branch of existing location theory for analysing urban hospital location would be central place theory. This was because the hierarchical and interlocking nature of the hospital system seemed, *a priori*, to fit the central place model quite well. Nevertheless, it was argued that certain extensions to the theory would need to be made in order for it to comply more with the complexity of the locational environment found in cities. A classification of hospitals was devised on which to base the empirical analysis in the second part of this research. This was designed to separate facilities into groups according to their degree of centrality which, basically, is an indicator of the extent to which they are oriented towards centres of population. As a result, it was argued that the type 1 facilities (acute hospitals) came closest to the requirements of central place approach. In the next three chapters the theoretical framework for analysing hospital locations in cities is developed. Chapter 2 considers the concept of districting in more detail and analyses the advantages and disadvantages of different types of districting criteria from the point of view of equity or efficiency. The role of an hierarchically organized set of services is then explained and the locational properties of hierarchical systems are analysed in detail. A set of methods is then introduced to show how the spatial organisation of hospital services is distorted by variations in population density for one particular districting criterion. The cost advantages of different systems of provision are then briefly analysed and compared. Chapter 3 examines the effects of differential accessibility in cities on the locations of hospital facilities. New districting patterns are developed based on equity of access where access is expressed in terms of travel time instead of distance. The methods are then applied to the special case of emergency medical provision in which speed of access is an essential requirement. In Chapter 4, attention turns to an analysis of the demand for health care services in cities. Following a discussion of the concepts involved, the spatial implications of different demand behaviour are examined in detail. Strategies are considered for modifying hospital services as population changes in each area of the city and some of the possible consequences are detailed. A planning model is then presented which allows the user to estimate the main effects of closing or opening facilities in different areas, in terms of patient flows, hospitalisation rates and catchment populations. Whereas some of the earlier analysis is conceptually based, this particular model is designed with practical considerations in mind. Chapter 5, the final chapter, evaluates the development of the London hospital system over an expanded time period. The main objectives are to consider how the sizes and locations of hospitals have changed in relation to the growth and de-concentration of population. In this respect the theory developed in earlier chapters is used as the basis for interpreting particular changes which have taken place.

2 The geographical organisation of hospitals in cities

2.1 Introduction

The method of dividing a region into areas each served by a hospital is defined as districting. In this chapter, the first of three concerned with the development of the theoretical aspects of the framework, different districting criteria are compared in terms of their spatial properties and then contrasted for their advantages and disadvantages. The three methods are chosen from a wide selection of possible criteria, partly because they cover a wide range of locational behaviour, but partly also because the key variables they use can be readily related to equity and efficiency objectives. They are interesting because under geographic conditions of uniform population density and travel costs they are equivalent to one another, but under other conditions they have a divergent behaviour.

Although the resulting patterns of provision may describe accurately many actual locational patterns, no arguments are advanced to suggest that the criteria be modified to fit these patterns exactly. The idea is that districting is useful for devising a set of ideal locational standards, against which comparisons can be made with actual locational patterns to see where and if there are particular divergences which require investigation or remedy. This is not to argue that districting is an essential characteristic of all types of hospital. Whether districting patterns can be recognised depends critically on the geographical scale of analysis. Within large cities, districting is most influential in the location of type 1 (acute) hospitals for reasons of accessibility; however, this is unlikely to be true of type 3 (psychiatric) hospitals. In small cities and at a very local scale, there may be no recognisable patterns at all. In the cases considered here, the most serious difficulty arises when hospitals are located very close together, as may occur in city centres, in which event the assignment of patients to particular hospitals becomes indeterminate. The intention therefore is to use the districting approach simply as an interpretive tool rather than as a means of prediction. This will be adequate provided that this and other complicating factors do not undermine the basic spatial principles of equitable access and efficiency of provision.

In developing the discussion, it would be impossible to deal independently with every possible shape, size or type of city. Every one is different. Rather, the approach is to select that type of city which in broad terms seems to be representative of the general form of many large cities in the world today. The cities in question are those having a broadly circular shape and a pattern of population density which is symmetrical with respect to the city centre and which shows a decline on any radius from the centre to the city perimeter. Although this is clearly an idealisation, it is a simplification which enables the discussion to proceed more smoothly than would otherwise be the case.

With the above introduction as background, the discussion begins with a general description of the three districting criteria. A simple framework is introduced for evaluating their properties in more detail, particularly in terms of their implications for patient accessibility, resource requirements and economy of provision. The concept of multilevel districting is then introduced to generate simple hierarchical patterns of hospital provision on the basis of central place theory. Methods are considered to enable different districting patterns to be altered to fit particular patterns of population density in different cities. Finally, at a general level, the problem of scale economies in hospital size is explored in relation to the districting criteria and the running costs of the hospital facilities provided.

2.2   Districting under even and uneven densities

The three main factors that must be taken into consideration when allocating a fixed quantity of health care resources to a region are the number, size and spacing of the hospitals. A large variety of economic profiles can be generated for different combinations of the three factors based on the cost structure of the resources involved. To simplify matters initially, the number of facilities is taken as fixed, and the prices of land and buildings are regarded as unique costs, so that the only factors that are assumed to vary are the size, spacing and running costs of the hospitals. Furthermore, it is assumed that the hospitals are free to locate anywhere, and that there are no barriers to travel in the city preventing access between different areas.

First, consider the problem of how far apart the hospitals are likely to be. It may be supposed that high accessibility costs will result in compact districts to minimise the inconvenience to patients, whereas with low costs of accessibility other factors, such as the total population, may be more important in determining the size of a district. As a basis for discussion, and for examining actual patterns of hospital location, three districting criteria are defined:

(1)   To divide the urban region such that the resident population dependent on a hospital is the same in each district (known as *P-districting* for short)
(2)   To divide a region such that the aggregate distance of travel of the population from a hospital is the same in each district (*TC-districting*)
(3)   To divide a region such that the maximum distance of travel of a resident in a district from the nearest hospital is the same in each district (*MC-districting*).

Many other criteria of varying use and complexity might have been chosen instead. For example, it is equally valid to determine districts so that the median distances of the populations from the hospital are the same. It should be made clear from the outset, therefore, that the objective is not to inquire which of all the criteria is best. It is rather to consider their implications given different patterns of population and accessibility costs, subject to a variety of assumptions concerning resource costs and city size. In the analysis the P and MC criteria prove the most useful for these purposes. Because they represent two extremes among a range of different districting options, it implies that other criteria will usually lean either towards one or the other in terms of their geographical properties.

2.3   One-dimensional urban regions

It is simplest to start the evaluation of these criteria using the abstraction of a one-dimensional urban region in which the population is spread continuously along a line. The results may then be extended to the more complicated but realistic case of a two-dimensional region. Figure 2.1 shows two one-dimensional regions with a variety of curves and lines added to convey important details about the properties of the criteria. The horizontal axis represents the distance from the city centre located at O (only the right-hand portion of the city is drawn); the vertical axis represents the population density, travel distance, and aggregate travel distance, and is scaled to convenient units. Both regions are divided into five equi-distant districts each served by a facility (L_1–L_5) located at the mid-point. Vertical lines (I, J, K) representing the district boundaries and meeting the horizontal axis at d have been drawn differently for each districting criterion, P, TC or MC, whereas the other lines convey key information about different characteristics of each district. Specifically, solid lines (e.g. B) show the linear increase in travel distance of an individual from the nearest facility; the curved, dotted lines (e.g. C) show the increase in aggregate travel distance of the resident population from the nearest hospital; and a thicker dotted line, denoted A, shows the variation in population density from the city centre O to the city periphery Q. In case (a) this line is horizontal, indicating a uniform or even density; in case (b), for the purposes of illustration, the line slopes linearly from left to right, indicating a non-uniform population density.

Examining both regions, it is seen that, in case (a), the district boundaries (I, J, K) coincide at the mid-point between each hospital, d. This point is the intersection for lines representing distance (B) and aggregate travel distance (C), but it also divides the resident population (A) on each side into equal amounts. Hence, it follows that all three criteria can be scaled to give the same result yielding identical location patterns. In case (b), where population density varies, this convenient property vanishes. Although district boundaries based

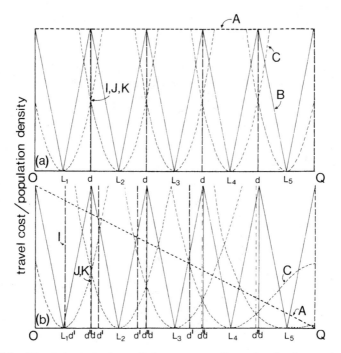

Figure 2.1 Districting criteria applied to two one-dimensional urban regions (a) with an evenly-distributed population, and (b) a linearly-declining population. -------, Density; —————, travel costs; – – – – –, total costs; — — —, P-divide; ——————, MC-divide; – · – ·, TC-divide. A, population density; B, travel costs; C, aggregate travel distance; I, P-divide; J, MC-divide; K, TC-divide; L, hospital location; d, district boundary case (a) and MC-criterion; d′, district boundary case (b) and P-criterion; d″, district boundary case (b) and TC-criterion.

on the MC criterion remain unchanged by definition, the district boundaries of the TC and P criteria fail to coincide by different amounts depending on the form of the density profile. For this particular example, it is specifically deduced that if the positions of the hospitals remain the same:

(1) The population served by each hospital must decrease from left to right.
(2) The aggregate distance of travel must also decrease from left to right because there are fewer people resident who might travel.
(3) The intersections of the district boundaries (located at d, d′, d″) under each criterion become increasingly dislocated from each other with distance from the city centre as density declines.

In the above case, the hospitals were fixed in position prior to the calculation of the population and aggregate travel distance; however, consideration of their locations in relation to the new boundaries raises further interesting questions. The purpose behind the approach taken, therefore, is simply to give a first illustration of some basic differences in districting properties. This is a problem which may be further refined and approached, less easily, from a variety of other standpoints. There is one unusual but just plausible case that is worth mentioning before proceeding. If the population density undulates regularly at even intervals and if the hospital locations match the rhythm of the undulations exactly, then again the criteria will coincide. Such occurrences should be regarded as special instances of the evenly populated case.

2.4 Resource allocation under different districting criteria

Although a variety of criteria may form the basis for allocating health care resources, the simplest and most convenient is the population of a district. More refined measures based on factors such as age, sex and morbidity may also be devised. The essential point is that, in a rational system, there ought to be a proportional relationship between hospital size and whatever measure is adopted. Because of the different spatial properties of the districting criteria, it seems clear that this will result in completely different patterns of provision. Only in the case of the P-criterion (where P is assumed to be the actual or weighted population) or in an evenly populated region would all the sizes of the hospitals be the same. For the TC and MC criteria, the hospital sizes would decline in areas of sparse population – in the case of the latter in direct proportion to the population density. This means that at the periphery of cities the hospitals would tend to be smaller. It is important therefore to make a case for each pattern of provision, and at this stage in the discussion, three general observations are possible.

Firstly, hospitals based on the P-criterion can be built to similar specifications to obtain economic advantages of scale. The resultant standardisation of services across the region would then facilitate administration and reduce running costs, since all hospital facilities would function similarly in terms of manning levels, use of equipment and throughput of patients. In this way the performance of individual hospitals could be carefully regulated. A disadvantage, however, would be the greatly increased distances that patients are forced to travel in sparsely populated areas. Secondly, resources allocated under the TC-criterion would achieve an improvement in accessibility in peripheral locations but there would be a corresponding reduction in hospital size. The conditions under which health authorities would want to consider such a scheme, however, are more obscure than for either of the other criteria. They appear to hinge on the question of who pays for the travel costs. If, for example, there is an extensive ambulance service or a travel cost reimbursement scheme in operation, then this criterion would clearly be advantageous. Thirdly, under the MC-criterion, the distribution of benefits between the providers and patients moves decisively to the latter, because no one is ever more than a fixed distance from a hospital. Such an arrangement is therefore closer to the philosophy of providing each local community with its own health care facilities. The basic problem, however, is that in sparsely populated areas hospitals would be very small, making it difficult to ensure an economic mix of services.

2.5   Allocation mechanisms in central place health systems

The determinants of hospital location in health care systems do not have the basic test of profitability which is the feature of private enterprise. In Chapter 1, it was argued that the principal goal of hospitals was to satisfy health needs. In central place theory, the profitable provision of services is dependent on only two mechanisms: the outer range and threshold of a service. For health services, it is thus necessary to re-interpret these mechanisms in the light of the differing objectives of health care systems. This will enable a formalisation of the relationships between districting principles and central place theory, and also prepare the ground for the more general analysis in later sections of the multilevel provision of health services.

Outer range
The outer range in central place theory defines the maximum distance a patient is prepared to travel to obtain a particular service. Unlike the MC-criterion which is a decision rule determined by health authorities, the outer range is a behavioural measure which varies according to the nature of the service, in particular how frequently it is needed. For example, routinely provided services are observed to have smaller ranges than specialised services. In practical terms, however, the concept is usually difficult to apply with consistency because there is always a small probability of some patients exceeding the range and because the measure is not independent of supply considerations (see Ch. 4). Hence, these limit its usefulness as a behavioural principle with which to delineate district boundaries. A more useful interpretation is to regard it either as a decision variable for locating different services or as an indicator of the quality of service in a locality. For example, in the case of P-districted hospitals, access standards cannot be uniformly applied, so that the range in sparsely populated areas is of necessity higher than in highly populated areas.

Thresholds
The second mechanism, the threshold, is defined as the level of demand needed for the provision of a particular service. Thresholds are high relative to demand for those services involving specialised skills and costly treatment, whereas they are low for routinely provided services. This means that the

former services tend to be highly centralised and the latter more dispersed. The absence of a profit criterion, however, is likely to result in an imperfect operation of the threshold because, as was seen in Chapter 1, there is no compelling financial incentive for hospitals to achieve a high degree of spatial efficiency. Historically, the stimuli to develop hospital facilities in new localities tended to be based on perceived needs but, because of the nature of health care systems, this approach leads to some arbitrariness in terms of locational decisions. In modern nationalised health care systems, thresholds may be formally imposed by a higher authority by linking them to demand, service priorities and strategic planning considerations and this can result in improvements. In a market-orientated health care system, some financial incentives exist based on satisfying demand, but they may be uneven in their impact so that irrational behaviour in hospital provision can and does occur. The key point is that the threshold, unlike its central place counterpart, is less efficient as a regulatory mechanism in health services than in the private sector. On the other hand, to ignore it would remove the only available benchmark for evaluating differences in services in different areas. The intention, therefore, is to conduct the theoretical analysis as if the threshold were an effective regulatory mechanism in the sense that it is representative of costs of, and demands for, different services. Whether it is effective in practice depends very much on the particular health system being studied.

Combining the threshold and range
Associated with the threshold in central place theory is a circular region around each hospital in which the demand generated is sufficiently large to justify the introduction of the service in question. Providing the outer range of the patients' willingness to travel is greater than or equal to the inner range of the threshold, then according to the theory that service ought always to be provided. If the inner range exceeds the outer range, then the threshold is not satisfied and the service is withdrawn. Note that this spatial interpretation of the threshold mechanism is appropriate only if an outer range exists. If the latter is indeterminate, as suggested earlier, such a precise geometric analogy will not be available. Nevertheless, the spatial representation is useful in the following sense. Suppose some hospitals – for instance in rural areas – are allowed to function at below threshold demand in order to maintain local services. Clearly, from an efficiency standpoint, better use of those resources could be made elsewhere and so, in this case, potential conflict arises. For cities in particular, it is of critical importance to set out the options arising from this conflict that could be used for striking a balance between, on the one hand, service-based considerations (i.e. the rural case) and, on the other, the need for locational efficiency. Broadly speaking, these options are as follows:

(1) The thresholds are strictly observed by health authorities and hospitals, but in very sparsely populated areas (i.e. on city perimeters) mobile

medical facilities or ambulance services are provided to ease the accessibility problem.

(2) Hospitals in low density areas provide fewer services than those in high density areas, but they are still relatively accessible to the population. For more specialised services, people travel to the city centre where it is easier to satisfy thresholds.

(3) *Either* hospitals in low density areas are equipped with more or less the same services but at the price of poorer accessibility, *or* accessibility is given priority and services are provided at lower than threshold levels. If the second option is taken then some areas and some services will effectively be subsidising others.

(4) Hospitals offer substitute services that are not the best available, but which are viable at lower thresholds. This means that the *quality* of the service provided in low density areas might also be lower as a result, but the services remain reasonably accessible.

2.6   A spatial hierarchy of urban hospitals

The strategy of providing certain services in some locations but not in others entails a hierarchical arrangement of hospital services and sizes. In the logic of central place theory, this leads to a very precise and well known set of spatial patterns. If we consider again an evenly populated region, the circular districts discussed hitherto should overlap with neighbouring districts supplying the same services until the urban region is filled without any gaps. The question then arises as to the minimum number of hospitals needed to achieve complete coverage of a city given a set of services, and service thresholds.

Christaller's approach (Christaller 1960) was to seek a way of tessellating the region such that each service served the maximum area for a given radius, while satisfying the constraints imposed by the thresholds and outer ranges which determine the particular services to be provided. It may be shown that any regular geometrical figure maximises this ratio, but only three – triangles, squares, and hexagons – do so without overlap. Of these, the hexagon has the largest ratio and is hence the most efficient since it requires the least number of locations to achieve the complete coverage of any area. (A mathematical note based on this argument is given at the end of Ch. 3)

As an example, Figure 2.2 shows a region which has been tessellated with regular hexagons and where there are five overlapping levels of service provision. These range from hospitals operating in the smallest districts providing services at the very local level, to one very large hospital located at the city centre providing services for most of the urban area. Neither the number of levels and hospitals shown in this example nor the way the hierarchy is arranged has any particular factual basis at present, although five levels are probably the maximum number of independently distinguishable

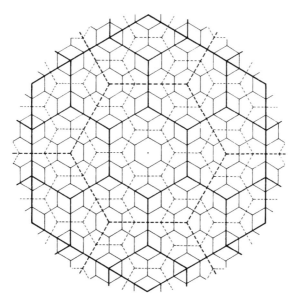

Figure 2.2   A five-level hierarchy of hospital services in a circular city region. District boundaries are represented by different widths and line qualities for each level. The largest hospital is located at the city centre.

levels that could be expected in a well organised system. Thus, in this case the central hospital would probably correspond to a large university teaching hospital with many beds, providing a whole range of services from the very general to the highly specialised. The intermediate levels, by contrast, would consist of various categories of general hospital, while the lowest level would typically represent community hospitals or clinics. In general, the exact terminology used to describe each type of hospital, the number of levels in the system, and their organisational basis will vary from country to country. In this respect, central place theory provides considerable flexibility as to the range of possibilities. For the moment, however, a number of general statements concerning the pattern observed can usefully be made.

(1) Higher level hospital facilities are less numerous, and provide more specialised services, but they also include the services offered at lower levels.
(2) The hospital at the centre of the city operates at the highest level, and has responsibility for providing specialised services for the entire population of the city as well as for services at the local level.
(3) Points inside each district are closer to the district facility than to any other facility of similar level, thus giving rise to the 'nearest supplier' rule.
(4) If the ranges of intermediate services are more than 0.91 times the radius of the district and if thresholds are strictly applied, then the services in

question may not be sustainable at that level and will move to a higher level. This argument is based on the radius of the circle having the same area as a regular hexagon. The service corresponding to this radius is termed the *critical service*.

(5) Some districts will necessarily share their services with areas outside the region; that is, for circular city regions, it is geometrically impossible with this arrangement to make the system self-contained.

(6) The allocation of resources per capita is always highest at the city centre and lowest at the city periphery. Equity therefore exists only to the extent permitted by the hierarchy.

It is interesting to note in passing that there is one sense in which this pattern could be improved upon. This is the case when health care services are delivered direct to the home, rather than patients having to travel to hospital. There are, of course, numerous economic and medical obstacles to this practice, if extensively used. Nevertheless, it should be regarded as the most dispersed and equitable pattern of provision, since evidently there are no thresholds acting which force services to coalesce in buildings and make patients travel.

From the viewpoint of the districting criteria, Figure 2.2 has another interesting interpretation in that the underlying population is assumed to be evenly distributed, so that the pattern of locations simultaneously satisfies the P, TC and MC criteria. However, if the population is neither uniform nor regular, then only the MC criterion is completely valid, in which case it is possible that only the hospital at the city centre will satisfy the service thresholds. If the health authorities wished to satisfy the thresholds at every hospital, P-districting would have to be the required pattern of provision. The problem then is how the tessellation might have to be distorted to meet the required thresholds and what implications this would have for the accessibility characteristics of individual districts. Before considering this problem, we complete the evaluation of the simpler geographic properties of the hierarchy.

2.7   Geographical properties of a simple hierarchical system

For an evenly populated, regularly tessellated region with the above hierarchical arrangement of facilities, there exist a number of well-known relationships between the area, population and travel characteristics in each level. Some of these relationships are now summarised in order to complete the discussion of the properties of the basic tessellated urban area. Figure 2.3 shows a portion of this region taken from Figure 2.2 in which there are six hospitals of level 5 (A–F) and one hospital of level 4 (O). The purpose of this section is to consider the spacing between these facilities, the radii and areas of the districts and the aggregate distance of the population within a district to the

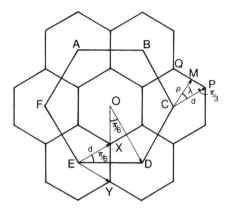

Figure 2.3 A portion of the five-level hierarchy showing six level 5 hospitals (A–F) and one level 4 hospital (O).

nearest hospital, and then to interpret these measures in the light of the discussion so far.

Spacing
If the radius EX of a district in level 5 is equal to d, then the district ED to the nearest hospital of similar level is given by $2d \cos \pi/6$ or $\sqrt{3}d$. By extension, the distance between two hospitals in level 4 is $2OD \cos \pi/6$ or $(\sqrt{3})^2 d$. In general, if the level of the hospital is m ($m = 1, \ldots, 5$), then the distance to the nearest hospital in the same level is given by $(\sqrt{3})^{6-m}d$.

District radius
If the radius of a district in level 5 is d, then the radius OD in level 4 is $\sqrt{3}d$. It follows that the radius of a district in the mth level is $(\sqrt{3})^{5-m}d$.

District area
The area v_5 of a district in the fifth level is found by determining the area of triangle YEX and multiplying the result by six. The area of this triangle is $(d^2/2) \cos \pi/6$ or $(d^2\sqrt{3})/4$, so that the required area is

$$v_5 = \frac{3}{2}\sqrt{3}d^2 \tag{2.1}$$

Repeating the procedure for other levels, the general expression for the area of a district in the mth level is found by extension to be $3^{5-m}v_5$.

Total distance
The total distance of the population in a district from the facility in the district τ is readily found using calculus, although the problem is slightly more complex.

Briefly consider triangle CPQ in Figure 2.3. If the aggregate travel distance of the population from the hospital located at C can be determined, then the result can be multiplied by six to give the equivalent result for the whole district. To proceed, it is first convenient to draw a line of length ρ from C to an arbitrary point M on PQ, representing one edge of the hexagon, and to let the angle formed with CP equal λ. This means that angle CMP becomes $(2\pi/3) - \lambda$. Next it is noted that, by the sine rule,

$$\rho = CP \sin\frac{\pi}{3} \bigg/ \sin\left(\frac{2\pi}{3} - \lambda\right),$$

which is equivalent to $(d\sqrt{3}/2)/\sin(2\pi/3 - \lambda)$. Now, the aggregate distance from C is obtained by integrating λ between 0 and $\pi/3$ along the hexagonal boundary, line PQ. This is equivalent to finding a solution to the following problem:

$$\frac{\tau}{6} = \bar{D} \int_0^{\pi/3} \int_0^{\frac{d\sqrt{3}}{2\sin(2\pi/3 - \lambda)}} \rho^2 \, d\rho d\lambda \tag{2.2}$$

where τ is the aggregate distance for the whole hexagon and $\bar{D}$ is the population density. Evaluation of equation (2.2) gives, after much algebra,

$$\frac{\tau}{6} = \frac{\bar{D}\sqrt{3}d^3}{8} \int_0^{\pi/3} \operatorname{cosec}^3\left(\frac{2\pi}{3} - \lambda\right) d\lambda \tag{2.3}$$

$$= \frac{\bar{D}\sqrt{3}d^3}{8} \left(2/3 + 1/2 \ln 3\right) \tag{2.4}$$

There are six triangles like CPQ in any district, so that τ actually equals $3\sqrt{3}\bar{D}d^3/4)(2/3 + 1/2 \ln 3)$. Replacing the second term in brackets by γ, which is constant for any regular hexagon, and generalising the result to all levels in the hierarchy, we then have τ_m equal to $(3\sqrt{3})^{5-m} (3\sqrt{3}\bar{D}d^3\gamma/4)$ or $(3\sqrt{3})^{5-m}\tau_5$.

Table 2.1 is a summary of the results derived in this section for the five-level system presented in Figure 2.2. As to its generality, three aspects are of interest to hospital planners. Firstly, the area, spacing and maximum radius are related to the geometry of the districts and do not depend on the density of the population, whereas the measures of population and total distance are density dependent. Secondly, if the population density is uneven, but the density distribution is known, it is always possible to derive a rough estimate of the population within a district provided that the district is small. However, this is not applicable to the derivation of the total distance which depends critically both on the size of the population and its distribution within the district concerned. This property makes this indicator less convenient in measurement terms.

Table 2.1 Elementary spatial measures in a five-level hexagonal hierarchy: the evenly populated case.

	Level	Area ($\times \bar{D}$ = population)	Spacing	Maximum radius	Total distance($\times \bar{D}$)
low	5	$\frac{3}{2}\sqrt{3}d^2$	$\sqrt{3}d$	d	$\frac{3}{4}\sqrt{3}d^3\gamma$
	4	$\frac{9}{2}\sqrt{3}d^2$	$3d$	$\sqrt{3}d$	$\frac{27}{4}d^3\gamma$
	3	$\frac{27}{2}\sqrt{3}d^2$	$3\sqrt{3}d$	$3d$	$\frac{81}{4}\sqrt{3}d^3\gamma$
	2	$\frac{81}{2}\sqrt{3}d^2$	$9d$	$3\sqrt{3}d$	$\frac{729}{4}d^3\gamma$
high	1	$\frac{243}{2}\sqrt{3}d^2$	$9\sqrt{3}d$	$9d$	$\frac{6561}{4}\sqrt{3}d^3\gamma$
multiplier		3	$\sqrt{3}$	$\sqrt{3}$	$3\sqrt{3}$

From a comparative standpoint, it is also noteworthy that total distance increases monotonically at a faster rate than the other measures, illustrating clearly how rapidly accessibility costs can increase as a consequence of concentrating hospitals in fewer locations. Finally, in general terms, it should also be emphasised that all the measures represent properties of the spatial system under consideration. They are unrelated to the more complex behavioural aspects of a patient's willingness to travel long distances to attend hospital. This is an issue which is taken up in Chapter 4.

2.8 Some simple transformation techniques for obtaining P-districted hospital facilities

To satisfy the service thresholds, it was suggested that health authorities would require a P-districted system of hospital facilities. However, in unevenly populated urban regions it would be impossible to retain the shape, area, contiguity and population of a district simultaneously. Thus a transformation is sought that preserves population and contiguity, but which changes the shapes and areas of the districts without significantly disrupting the balance of provision in any part of the city. There are a number of methods that can be used to develop suitable transformations, although in complicated cases complex technical issues can be involved. No attempt is made, therefore, to provide a complete treatment of the subject as this would lead to discussion too far from the issue of hospital provision. In the case of radially symmetric cities suitable transformations are reasonably straightforward to derive and a flexible technique can be developed. The discussion begins with a few general statements followed by a series of examples. The question of how to develop suitable transformations for more complicated urban regions is then briefly considered in general terms.

As a basis for discussion, it is useful to start by defining the function $F(\cdot)$, $0 \leqslant F(\cdot) \leqslant 1$, which is the proportion of the population contained in a region R of a city, or its image region R′. Specifically,

$$F(r, \theta) = \frac{1}{P_{r\theta}} \iint_R D(r, \theta)r \; dr \; d\theta \qquad (2.5)$$

and

$$F_1(u, v) = \frac{1}{P_{uv}} \iint_{R'} \phi(u, v)u \; du \; dv \qquad (2.6)$$

where $D(r, \theta)$ is the population density at coordinates (r, θ) in R, $\phi(u, v)$ is the density at coordinates (u, v) in R′, and $P_{r\theta}$ and P_{uv} are the populations of R and R′, respectively. The populations P act as convenient normalising factors, but are not essential to the current analysis if, as is presumed, the populations of both regions are the same. Also, if the density in either region is assumed to be symmetric, then $F(\cdot)$ is independent of θ and v, and so these may be omitted.

Finally, it is usual to compare equivalent sectors of the urban areas in R and R′. In the examples presented, the sectors are considered to be completely circular areas and so θ and v range over 2π radians; however, with very minor modifications to the technique any size of sector may be analysed. With these conditions in mind, we begin by equating a proportion of the population in one region with an equal proportion in the other;

$$F_1(u) = F(r) \qquad (2.7)$$

Rearranging Equation (2.7) and making r the subject, we may then write

$$r = F^{-1}[F_1(u)] \qquad (2.8)$$

This relationship is fundamental and is said to define the transformation from region R′ to region R, whereas the opposite relationship,

$$u = F_1^{-1}[F(r)] \qquad (2.9)$$

in which u becomes the subject of the equation, defines the inverse transformation from region R to region R′.

Let the population density in R′ be constant and equal to $\bar{D}$, and let the radius be u_{max}, then

$$F_1(u) = \frac{\bar{D}}{P_u} \int_0^{2\pi} \int_0^u u \; du \; dv \qquad (2.10)$$

For a constant density, the population of R' is $\bar{D}\pi u^2_{max}$, so that on integration and cancelling terms, Equation (2.10) simplifies to

$$F_1(u) = \frac{u^2}{u^2_{max}} \qquad (2.11)$$

From Equations (2.9) and (2.11) we have, therefore,

$$u = u_{max}[F(r)]^{1/2} \qquad (2.12)$$

which is the inverse form specified in Equation (2.9).

2.9 Examples of transformations

The above results are now used in a sequence of illustrative transformations. When the direction of the transformation is from R' to R, where R' is evenly populated, the result turns out to be an irregular tessellation of six-sided, curvilinear polygons (where R is also evenly populated, then the identity transformation results, see Example 1). These polygons locally approximate the regular hexagons found in R', but they are distorted according to the area of the district and the form of the density function, $D(r)$. Every polygon contains an equivalent proportion of population to its image district, but both TC and MC properties are forfeited as a result of the procedure.

Example 1 (A uniformly distributed population)   This transformation maps districts in R' on to a region R in which the population is also evenly distributed. In region R, the proportion of the total population contained within an area defined by r, the distance from the city centre, is

$$F(r) = \frac{r^2}{r^2_{max}} \qquad (2.13)$$

Where r_{max} is the maximum urban radius. Equating $F_1(u)$ with $F(r)$, rearranging and taking the square root, we have

$$r = r_{max}[F_1(u)]^{1/2} \qquad (2.14)$$

If $r_{max} = u_{max}$, then it follows that

$$r = u \qquad (2.15)$$

Since the tessellation would be unchanged in this example, the result is known as the identity transformation.

Example 2 (A normally distributed population)   A possible form of $D(r)$ for a city is the normal distribution, which may be written

$$D(r) = A \, e^{-br^2}, \tag{2.16}$$

where A is the density at the city centre and b is the exponential rate of density decline. To proceed with the transformation, it is first noted that the population of R is given by

$$P_R = 2\pi \int_0^\infty A e^{-br^2} r \, dr \tag{2.17}$$

$$= \frac{\pi A}{b} \tag{2.18}$$

The proportion of the total population contained within r is then

$$F(r) = 2b \int_0^\infty e^{-br^2} r \, dr \tag{2.19}$$

$$= (1 - e^{-br^2}) \tag{2.20}$$

Equating $F_1(u)$ with $F(r)$, we have

$$F_1(u) = (1 - e^{-br^2}), \tag{2.21}$$

so that

$$r = \{-(1/b) \ln[1 - F_1(u)]\}^{1/2} \tag{2.22}$$

which is the required result. Figure 2.4 shows the resulting transformation based on Equation (2.22), in which a parameter value of b has been used corresponding to a regression estimate derived from the London pattern of population density in 1971. The transformation illustrates three general principles firstly, the curvilinear shape of the hospital districts becomes more pronounced at higher levels in the hierarchy. Secondly, individual districts become increasingly elongated with increasing distance from the city centre. Thirdly, the locations of all hospital facilities, except the centrally located hospital, are dislocated so that the 'nearest centre' feature of the standard central place tessellation holds exactly only at the city centre.

    In carrying out such a transformation, it should be stressed that the degree of spatial distortion to districts (e.g. lateral or radial stretching) is dependent on the conditions imposed by the analyst, whereas the choice of the right conditions for a particular city is essentially an empirical question. In this and

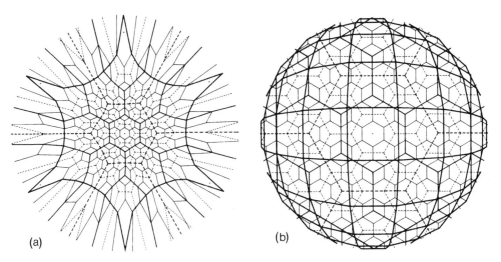

(a) (b)

Figure 2.4 The normal (a) and inverse normal (b) transformations showing a five-level hierarchy of hospitals in a city. Districts are distorted to retain the same population in each district.

the subsequent cases, the elongation of hospital districts in the suburbs would be consistent with the orientation of transportation routes. Later, in Chapter 4, survey data are used as a check on this stretching property. Another noteworthy aspect of this particular example is the way the urban boundary has been defined. The normal distribution is asymptotic to r so that there is no natural choice of urban radius. In the present case it has been arbitrarily set as the radius of a circle along whose circumference the average population density is one person per unit area.

The inverse of Example 2 is the transformation from R to R′, and this is simply

$$u = u_{max} (1 - e^{-br^2})^{1/2} \tag{2.23}$$

In Figure 2.4b, the inverse is shown as an initially square grid which has been distorted and overlaid on the image region R′.

Example 3 (A linear population density) A form of density function used at the beginning of this chapter showed population density as declining linearly between the city centre and urban boundary (see Fig. 2.1b). This can be written as

$$D(r) = A - br \tag{2.24}$$

where b is the linear rate of density decline. Firstly, it is noted that the population of R is

$$P_R = 2\pi \int_0^{r_{max}} (A - br)r \, dr \qquad (2.25)$$

where $r_{max} = A/b$. Integration of this equation gives

$$P_R = (\pi/3)Ar_{max}^2 \qquad (2.26)$$

We therefore have

$$F(r) = \frac{6}{r_{max}^2} \int_0^r r - \frac{br^2}{A} dr \qquad (2.27)$$

$$= \frac{3r^2}{r_{max}^2} - \frac{2r^3}{r_{max}^3} \qquad (2.28)$$

After equating $F(r)$ with $F_1(u)$, in the usual way, the solution for r is obtained from the following polynomial equation

$$0 = 3r^2 r_{max} - 2r^3 - F_1(u)r_{max}^3 \qquad (2.29)$$

The resulting transformation is shown in Figure 2.5, which has similar features to Example 3 except that the stretching of districts is less pronounced.

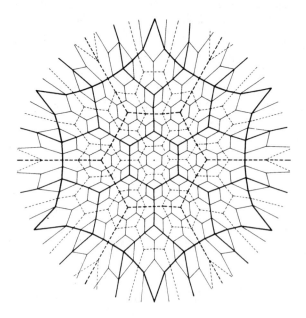

Figure 2.5   A P-districted city with a linearly declining population density.

Example 4 (A negative exponential population density) A typical and extensively applied form of urban density function is the negative exponential, which will be further analysed in Chapter 5. It is written as

$$D(r) = A e^{-br} \tag{2.30}$$

where b is the exponential rate of density decline. The transformation from R′ to R cannot be obtained analytically, but may be graphed or solved iteratively. The inverse transformation from R to R′, however, is readily found. As before, we first find the total population P_R, which is given by

$$P_R = 2\pi \int_0^\infty A e^{-br} r \, dr \tag{2.31}$$

Integration by parts gives

$$P_R = 2\pi A / b^2 \tag{2.32}$$

Similarly,

$$F(r) = 1 - (1 + br)e^{-br} \tag{2.33}$$

Substituting (2.33) in (2.12), it is easily seen that

$$u = u_{max}[1 - (1 + br)e^{-br}]^{1/2} \tag{2.34}$$

which is the required inverse.

The above description and examples of a suitable technique for determining P-districted systems of hospital provision takes the discussion as far as is necessary for the problems considered in this chapter. It suffices to say that the method can be modified, restated or adapted in a number of ways, but the results add relatively little to the conclusions already apparent. The problem becomes interesting again in the case of cities which either do not possess radial symmetry or which belong to a completely different category of city. Tobler (1963) has considered the general form of the equations needed and the sorts of considerations and difficulties that might occur in these cases. The subject of transformations is not considered again until Chapter 3 when the basis for the transformations is not population density, but accessibility patterns.

2.10 Scale economies, hospital size and districting

The last stage of the discussion is to compare more carefully the cost implications of selecting one mode of districting rather than another. Thus far, it has been established that MC-districting is likely to be more equitable for patients in sparsely populated areas but not particularly efficient. Conversely, P-districting is likely to be more efficient but not very equitable. The questions considered in this section are designed to look closely at the basis of these conclusions in order to establish the conditions of their validity.

The first requirement is that there should be economies of scale from building larger rather than smaller hospitals. Whether or not scale economies exist depends critically on the component costs of the resources used. Economies tend to arise through the shared use of medical and other facilities, such as operating theatres, or X-ray units, pathology laboratories, laundries, canteens, manpower (doctors, nurses and ancillary staff). For a given quality of treatment and mix of patients, the case for introducing scale economies by building larger hospitals is thus fairly clear. A particular problem, however, is that quality and mix vary considerably and this has tended to frustrate efforts to establish the true value of the 'optimum' size for individual hospitals.

When data on hospital size over a lengthy period are grouped and examined, shifts are observed not only in the average sizes of the hospitals, but also in their size distribution, suggesting that the problem is even more complex than has been supposed. In modern health care systems, medical techniques have grown more sophisticated, relying on expensive equipment and specialist services. In older or developing health care systems, techniques are more rudimentary, suggesting that returns to scale are less attractive. Thus, it is of interest to speculate whether, over a period of time, the increase in the cost and

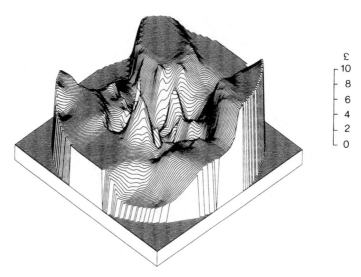

Figure 2.6 Variations in the cost per bed in the London area in 1921 within 50 km of the city centre.
Source: Burdett 1889–1930.

sophistication of medical care might have had any impact on actual patterns of hospital location in cities.

As a brief illustration of this point, it is instructive to turn to Figure 2.6, which shows a three-dimensional surface representing the hospital costs per bed in London and the surrounding area for acute hospitals in 1921. This year is of particular interest because, to some extent, the pattern of hospital locations at that time resembled the MC-criterion (see Ch. 5). A general trend is apparent with costs (except those near the city centre) steadily rising in most directions reaching a peak at the edge of the region. After 1921, hospital patterns changed more towards a P-districted pattern, and so a key part of the following analysis is to establish in more detail the possible cost factors involved in this transition.

Because a detailed analysis is potentially very complex, it is necessary to proceed with the aid of some simplifying assumptions. The first assumption is to consider the urban hospital system as if it contained only one level of provision rather than a strict hierarchy in the sense used earlier. The more complex question of size relativities *between* different levels is examined in Section 2.11. The second simplification is to consider the cost structure of the system at a very general level of detail, so that such questions as treatment quality and patient mix are disregarded and are presumed to be independent of location. To commence the analysis the following definitions are first established.

Definitions
(1) $A(r)$ is the area of a 1 km wide ring at a distance r from the city centre $(2\pi r \times 1)$.
(2) $D(r)$ is the density of population (as before).
(3) d is the radius of a district (as before).
(4) $v(d)$ is the area of a hospital district $(=(3/2)\sqrt{3}d^2$, see Sec. 2.7) based on the MC-criterion.
(5) P_n is the population of a district based on the P-criterion.
(6) ω is the desired level of provision in beds per head of population.
(7) $N(r \mid d)$ is the *smoothed* number of hospitals at distance r given a district radius d, where $N(r \mid d) = A(r)/v(d)$.
(8) $N(r \mid P_n)$ is the smoothed number of hospitals at distance r given a district population of P_n, where $N(r \mid P_n) = A(r)D(r)/P_n$, which in turn is $2\pi D(r)r/P_n$, the density of a ring multiplied by the area and divided by the district population, P_n.
(9) $C(r \mid d, \omega)$ is the number of beds per hospital at distance r given a district radius d and level of provision ω, where $C(r \mid d, \omega) = A(r) D(r)\omega/N(r \mid d)$, which is, from Definition 7, $v(d)D(r)\omega$.
(10) C_n is a constant which is the bed capacity of a hospital under the P-criterion, where $C_n = P_n \omega$.

(11) $\rho[C(r \mid d, \omega)]$ is the cost per bed for a facility of size C whose value depends on r, given d and ω.

(12) $\rho[(C_n \mid P_n, \omega)]$ is the cost per bed for a facility of size C_n, given P_n and ω.

The above variables are representative of the key factors used to show the dependency of hospital costs and locational patterns. Only Definitions 1 and 2 can be considered exogenous in the sense that they describe the characteristics of the region in which the resources are to be allocated. Definitions 3–10, by contrast, are all decision variables because they are related directly to the allocation process. The bed capacities of the hospitals represent the size variable C; they are determined by the populations of the districts and the quantity of beds allocated per head of population. Note that they need to be calculated differently for each districting criterion. For P-districted hospitals, size is constant throughout the urban area and is given simply by $P_n \times \omega$, where ω is the desired level of provision. For MC-districted hospitals, size is a function of d, the radius of the hospital district, ω, and the distance of the hospital from the city centre. On any ring of constant radius the size is equal, but the number of hospitals on each ring is proportional to the circumference.

The problem becomes more interesting with the introduction of Definitions 11 and 12, which introduce scale economies into the system. The presumption is that the cost per bed is related to C, the bed capacity of the hospital. Although it is not important here to establish the precise nature of this relationship, it is usual to assume that as C increases there are increasing returns to scale initially, so that ρ, the cost per bed, declines. However, as C increases further the returns to scale become negative and ρ begins to increase. The next task then is to establish the total costs of provision in a city according to each districting criterion using these definitions.

Derivation of the total costs of provision
Consider an area partitioned by the MC-criterion. The costs of provision at distance r are the cost per bed ρ multiplied by the number of hospitals N and the bed capacity C,

$$\xi_1(r) = N(r \mid d) \, C(r \mid d, \omega)\rho \, [C(r \mid d, \omega)] \tag{2.35}$$

Integrating Equation (2.35) gives the total costs of provision for the city under the MC-criterion:

$$\xi_1^* = \int_0^{r\max} N(r \mid d) \, C(r \mid d, \omega) \, \rho[C(r \mid d, \omega)]dr \tag{2.36}$$

$$= 2\pi\omega \int_0^{r\max} D(r) \, \rho[C(r \mid d, \omega)]r \, dr \tag{2.37}$$

where Definitions 9 and 1 have been used.

Turning now to the P-criterion, the costs of provision at distance r are, from Definitions 8, 10 and 12,

$$\xi_2 = 2\pi\omega D(r)r\rho[C_n \mid P_n, \omega)]$$ (2.38)

giving, on integration,

$$\xi_2^* = 2\pi\omega\rho[C_n \mid P_n, \omega)]\int_0^{r_{max}} D(r)r \; dr$$ (2.39)

which is the total costs of provision in the city under the P-criterion.

If the mathematical forms of Equations (2.37) and (2.39) are compared, it is plain that the total costs of provision depend on several factors, so that without being more specific it is not possible at this stage to say which is the cheapest mode overall. There are, however, several special cases that can be safely eliminated. Firstly, if there are no returns to scale and ρ, the cost per bed is constant, ξ_1^* will equal ξ_2^*. Since MC-districting gives a better access to patients, it is more equitable than P-districting, so that this would always be the preferred pattern of provision. Secondly, when the population density is even, we have, from Definitions 9 and 10,

$$C = v\bar{D}\omega = P_n\omega = C_n$$ (2.40)

Thus, ρ becomes a constant in Equation (2.37) so that again ξ_1^* and ξ_2^*, the total costs of provision, are equal under each criterion. Thirdly, it is seen from Equation (2.39) that, other things being equal, ξ_2^* will be minimized when $\rho(C_n \mid P_n, \omega)$ is a minimum. This occurs when hospitals are built at their optimum capacity, that is when maximum advantage is made of scale economies. It is noteworthy that this is not possible under the MC-criterion as the constraint on accessibility causes hospital size to vary, so that it will be fortuitous if any hospitals are built to the optimum size. A fourth case arises when there are ever-increasing returns to scale, so that it pays the provider to build a hospital as large as possible. From Equation (2.39) it can be seen that ξ_2^* is minimized when C_n, the capacity of the hospital, is a maximum, that is when all provision is allocated to one very large hospital at the centre of the city.

Further illustrations of the MC-criterion
In the case of the MC-criterion, it is instructive to develop the above discussion one stage further by examining how specific factors influence spatial variations in the costs of hospital provision. The most convenient way to do this is with the aid of the set of graphs shown in Figure 2.7. These illustrate a range of typical interrelationships between the variables shown in the earlier list of definitions.

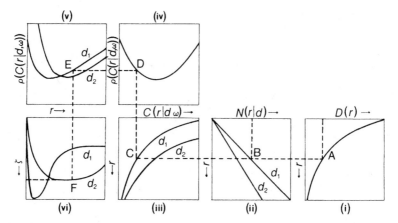

Figure 2.7   The graphs show: (i) the population density on distance from the centre r; (ii) the number of hospitals N on r; (iii) the capacity of a hospital located distance r from the centre; (iv) the cost per bed as a function of facility size; (v) the cost per bed as a function of r; (vi) total costs of provision, ξ_1, as a function of r.

Each graph links together with the adjacent graph. Graph (i) shows the relationship between population density, $D(r)$, and the distance from the city centre, r. Graph (ii) shows the increase in the number of hospitals as a function of r given a district radius of either d_1 or d_2 (where $d_2 > d_1$). Graph (iii) shows the resultant decline in hospital size as a consequence of the MC-criterion. Graph (iv) shows the relationship between the cost per bed ρ, given d and ω, the allocation of beds per capita. Graph (v) shows the consequent impact on the cost per bed for different values of r. Last, graph (vi) indicates how the total costs of provision vary with r.

To interpret the graphs, briefly consider a point A (graph i) some distance from the city centre. This point determines point B, the number of hospitals for a given district radius d_1, point C the capacity of the hospitals (given d_1 and ω), point D the cost per bed, point E the same cost, but expressed as a function of r, and point F, the total cost of provision at all hospitals located at that radius. When there are scale economies, as shown in graph (iv), the implication is that the cost per bed will vary according to r for different values of d. On the basis of further investigations, it was typically found that the costs at the city centre tended to be high. Farther out costs decreased as hospital sizes approached the optimum, and then increased again at the edge of the city. Similarly, the total costs of provision were high near the centre, but then fell before increasing again and levelling off at the edge of the city. Thus, cost variations are a strong feature of the MC-criterion; although these variations are strongly dependent on several factors, the net result based on reasonable assumptions points to higher cost facilities at the edge of the region.

2.11   *Hospital sizes in different levels of the hierarchy*

If the arguments of the previous sections are applied, not to one level in the hierarchy but to all levels, it is plain that it would be impossible to provide a system in which all the hospitals operated at the optimum size. It would, however, be interesting to test the sensitivity of the total costs of provision under different districting patterns and different assumptions concerning, say, the relative sizes of hospitals in each level. However, this would be a major analytical exercise. Fortunately, it is reasonably straightforward to examine simpler issues concerning the relative merits of either dispersing resources fairly evenly among locations or concentrating them on a few very large centres linked by much smaller satellite hospitals. This can be done in a large variety of ways using central place theory and this is a powerful feature of the approach. The following example is, therefore, used by way of illustration of what would need to be a more intensive and detailed analysis for use in particular cases.

The multiplier

The first steps are to decide the number of hospitals, the number of levels in the system, and the desired districting pattern. For simplicity, a P-districted city is selected here because this does not have the complication of size being dependent on location. In this respect, Figure 2.4, a P-districted city based on London, is thus a useful starting point. As with other examples given in this chapter, this particular case has five levels and a hierarchical multiplier of 3. This means that the cumulative number of hospitals in successive levels follows a geometrical progression, namely a, $a3$, $a3^2$, $a3^3$, $a3^4$, where a is the number of hospitals in level 1 (in this case $a = 1$). Although other multipliers could equally have been used (see Beavon 1977), the geometry of central place theory places a restriction on the values the multiplier can take. Christaller (1960), for instance, considered systems in which the multiplier took values of 3, 4 and 7, and these are probably the most practical cases as far as health care systems are concerned. Table 2.2 sets out the effect of the multiplier, denoted by l, on the numbers of hospitals in each level in general terms. The foot of column 2 shows that the total number of hospitals N is given by $al^{(m-1)}$, which can be obtained as the sum of a geometrical progression.

Some practical observations

In any practical exercise based on counting the number of hospitals in a city, an important distinction must be drawn between those hospitals serving districts which are wholly inside the urban area and those near the urban boundary which share their districts with areas outside. If a count is made of the hospitals actually located inside the city (the number increases by six on each ring around the centre), the total differs slightly from that determined by dividing the area of the city by the area of the smallest district in level 5, because the

Table 2.2 Hierarchical relations in a simple central place system.[*]

Level	Cumulative number of hospitals	Number of hospitals within each level
1	a	a
2	al	$a(l-1)$
3	al^2	$al(l-1)$
4	al^3	$al^2(l-1)$
.	.	.
.	.	.
.	.	.
m	$al^{(m-1)}$	$al^{(m-2)}(l-1)$
Total	$al^{(m-1)}$	$al^{(m-1)}$

[*]The smallest valid values of l are 3, 4, 7. For examples of other values and their spatial patterns see Beavon (1977).

Table 2.3 The number of hospitals and populations served in an l^3, five-level hierarchy.

Level	Number of facilities serving the city	Number of facilities located in the city	District population $\times 10^6$
1	1.21	1	6.86
2	2.42	6	2.29
3	7.26	6	0.76
4	21.78	24	0.25
5	65.35	60	0.08
Total	98.02	97	

Population of city $= 8.3 \times 10^6$; Area $= 4067$ km^2; Density function is $D(r) = A\,e^{-br^2}$, where $A = 92.76$ persons ha^{-1}; $b = 0.0035$ km^{-1}; $r_{max} = 35.98$ km.

latter procedure has a smoothing effect. Table 2.3 illustrates the difference in this case: column 1 shows the predicted number based on the multiplier rule above; column 2 shows the actual numbers. If the population of the city is taken as 8.3 million, as determined by the procedure for London shown at the bottom of the table and assuming a normally distributed population, then the population served by each hospital in any level can also be determined. This is given in column three.

Hospital size
Having established the total number of hospitals, in this case 97, located inside the city, the question now arises as to the size of each hospital.

There seem to be two basic options for proceeding. One is to evaluate the detailed economics of the system, in which case the sort of analysis carried out in Section 2.10 would be appropriate. The second, which is considered here, is to employ the existing stock of hospitals as a basis for establishing the sizing relationships which, though not optimal, provide results that are easier to build on. Although the following equation is by no means a unique expression of this relationship, it is one that has found wide application in other fields and is related to the familiar rank-size rule, that is,

$$C_i = C_1/i^\lambda \qquad (2.41)$$

In the current context, this equation is taken to mean that the size of a hospital C_i in the ith level is dependent on the size of the largest hospital C_1, but is inversely related to the level i and scale parameter γ. Thus, if C_1 were 1000 and γ were 1.0, the bed capacities of lower order hospitals in a five-level system would be 500, 333, 250 and 200 respectively. The idea is that if the values of particular variables were known, or could be estimated, then C_i can be calculated. Suppose, for example, bed availability in a city is given by C^*, and the number of hospitals in level 1 is a then C_1 can be determined from the following equation, which arises from a combination of the multiplier rule in Table 2.2 and Equation (2.4):

$$C^* = aC_1 \left[1 + (l - 1) \sum_{i=2}^{i=m} \frac{l^{i-2}}{i^\lambda} \right] \qquad (2.42)$$

An illustrative example
Briefly consider a situation in which the intention is to establish a provision of five beds per thousand at five different levels of hospital. The total number of beds required for a population of 8.3 million (see Table 2.3) is therefore 415 000. Application of the above steps for three different values of γ gave the results shown in Table 2.4. For each value of λ, there are two columns showing the size of the hospital and the total number of beds allocated to each level for use by the urban population. A third column in parenthesis shows the bed allocations in each level based on hospitals located inside the urban boundary. The difference between the second and third column gives the net number of beds that are shared with the region outside. Thus, in level 1 ($\lambda = 0.5$), about 184 beds nominally set aside for people living inside the city would be externally located, in level 2 much more of the responsibility is assumed by externally located hospitals. In level 4, about 973 beds inside the city would be nominally set aside for people living outside the city.

The results show interesting variations in the degree of concentration or dispersal that occurs as a result of changes in λ. High values have a concentration effect, whereas low values have a dispersing effect. When

Table 2.4 Variations in hospital size in a five-level hierarchy according to the ranking formula.

Level	$\lambda = 0.5$			$\lambda = 1.0$			$\lambda = 2.0$		
	Size	Beds in level		Size	Beds in level		Size	Beds in level	
1	879	1 063	(879)	1 777	2 150	(1 777)	6 291	7 612	6 291
2	621	1 504	(3 728)	889	2 150	(5 331)	1 573	3 806	(9 436)
3	507	3 683	(3 044)	592	4 300	(3 554)	699	5 075	(4 194)
4	439	9 570	(10 543)	444	9 677	(10 663)	393	8 564	(9 437)
5	393	25 681	(23 578)	355	23 225	(21 324)	252	6 442	(15 096)

$\lambda = 0.5$, for instance, a set of allocations is obtained with similarities to that found in London; when $\lambda = 1.0$, more resources are concentrated at level 1 and the result has similarities with the situation in a city like Vienna, Austria. Usually, 2000–3000 beds is, internationally, about the upper limit in size for acute hospitals, so that the case $\lambda = 2.0$ would clearly be unrealistic. Particular reasons for varying degrees of concentration in different cases is plainly an empirical question. From an analysis of London data (see also Ch. 5), considerable shifts in the degree of concentration are apparent over time and these seem to be related to two main factors. The first is the deconcentration of population which has had a reducing effect on the sizes of the largest central hospitals. The second factor rests on an economic argument involving the threshold mechanism. This says that, as resources for health care become more plentiful, some formerly high order services will be devolved to smaller centres to make them more generally available. Clearly, there is much experimentation that can be done with these mechanisms in testing different hypotheses, but if they are to be useful they must also take account of the preferences of patients for treatment in different locations. This issue is addressed in Chapter 4.

Conclusions

This chapter has been concerned with the problem of districting urban regions according to three types of criteria: equal population P, equal aggregate distance TC, and equal maximum distance MC. Hospital provision with differing degrees of health service specialisation was then dealt with, and conclusions were drawn regarding the conflicting aspects of service provision in terms of equity or efficiency. To obtain P-districts in unevenly populated urban regions a method of transformations was developed. The examples shown provided useful insights into the way in which a hierarchical system of hospital services may be geographically organised in a typical urban region. Building on these results, the problem of hospital costs was analysed and the spatial consequences of variations in scale economies were noted. The main objective was to establish a framework within which past and future patterns of provision could be gauged. This framework provided information on the geographical organisation of services but it did not deal with the question of shifting population and changes in demand behaviour.

The results indicated that important economic and social consequences depend on how hospitals are located in relation to each other and to the population they serve. In particular, a more efficient pattern of services is usually derived from an allocation of resources which is based on the population served by each hospital being equal. One important penalty for adopting this approach is a potentially reduced accessibility to hospital services in areas of low population density. Further economies (but also compensating reductions in accessibility) may be derived from the concentration of some services in strategic locations, particularly those highly specialised services which consume large quantities of resources in relation to the number of patients they treat. The extensive specialisation of activities leads ultimately to an hierarchical pattern of services of which many examples were given. Too great a concentration of resources, it was argued, might (but not necessarily always) lead to greater economies, but it would also cut the use made of those services by potential patients living far from the facility because distance has a deterrent effect on demand. The deterrence of distance is thus an important consideration in the locating of health care services, but its effects depend on individual circumstances and the service in question. These are issues which are considered again in Chapter 4. Before then we turn to a more detailed examination of the measurement of accessibility and consider in some depth what influence this may have on the location of certain hospital services.

3   The impact of travel time on the accessibility of hospitals in cities

3.1   Introduction

The accessibility of health care services formed a basis for the earlier development of hospital districting patterns. If attention is focused on finding an adequate measure of accessibility, however, it is apparent that a number of potential difficulties have been ignored. Patients pay for accessibility costs in terms not only of money but also in time, discomfort and many other factors. Distance has been used to approximate these costs, but it will be inadequate if the ease of movement depends largely on the variable character and efficiency of the transportation services in a city. The aim of this chapter, the second considering theoretical issues, is to examine the potential impact of variations in accessibility on hospital location and the districting patterns developed in Chapter 2. In particular the objective is to re-evaluate some of the problems considered hitherto using a measure of accessibility based on journey time rather than distance. Introducing this measure, however, raises numerous problems, because it is known that travel time varies on different parts of a transport network according to the mode of travel, level of congestion, time of day, weather conditions and many other factors. Furthermore, over longer periods it is also affected by changes in transportation technology and efficiency. A particularly difficult issue, for example, is whether individuals are able to make use of the best or most convenient modes of travel which in turn is related to factors such as income, age, transport availability and so forth. Thus, at a very detailed level of analysis, the problems entailed are exceedingly complex.

In order to make significant progress, therefore, the first step is to make certain simplifications. Rather than dealing with details of specific journeys, it is sufficient for current purposes to base the analysis on estimates of the local average speeds of movement as they vary across a city. Functions describing such variations, called *velocity fields*, have been extensively studied by Angel and Hyman (1976) and by changing the form and parameters of the functions concerned, it is possible to simulate in a very simple way a variety of known and presumed variations in travel efficiency and hence travel time. The most important assumption underlying the use of the velocity field approach is to presume that travel costs, measured in time per unit of distance, are far higher in central areas of a city than in outer areas (that is the average speeds are lower). The extent of this cost or speed differential is highly dependent on the form of the city. Those cities with old, historical centres and narrow, twisting streets are more at risk than modern, extensive cities which are usually designed with the car in mind. Indeed, it has been the relative growth in car ownership and travel as compared with the use of public transport which underpins the need to consider short and long run effects of changing travel times.

Whereas public transport, historically, has tended to emphasise access to and travel within the central areas of cities, and therefore the use of centrally located hospitals, car travel promotes the attractiveness of other, non-central locations. For example, it is possible to show, using the approach described here, that the locus of the most efficient and accessible locations in terms of car travel has shifted from the centre of London to a ring between 6 and 10 km out. Plainly, this will have a destabilising effect on certain centrally located hospital services, and there are now several examples of hospitals which have changed location in recent years because of reduced accessibility, smaller catchment populations and other contributory factors. These changes, in turn, raise numerous issues regarding locational equity versus efficiency which depend partly on the size of car-ownership and the distribution of income among hospital users.

A more related but complicated set of spatial effects derives from the fact that locations which seem to be geographically very near in distance terms may be very far apart as measured by travel time. Using public transport, the focal nature of the routes often obliges users to break their journey at the city centre because it is quicker than taking a more direct route. Similar journeys undertaken by car, however, rely more on circumferential routes to avoid delays caused by congestion. Clearly, therefore, there will also exist a so-called modal-split effect because, depending on the relative availability and efficiency of public transport, car owners have the possibility of choosing their level of access and trading it off against the quality of service they expect to receive. This enables them to select the hospital service they require from a far wider net of spatial alternatives. These fundamental differences, taken in isolation, seem trivial but considered in the light of all the journeys undertaken and in conjunction with other trends and factors, such as travel preferences and population deconcentration, the spatial and locational ramifications are profound.

Most of the aims of this chapter may be achieved by dealing with one particular branch of the health service, whose operation hinges critically on the time taken to gain access to different locations. This is the question of providing accident and emergency (A & E) services where there is a clear need to concentrate resources (ambulance depots, and treatment centres) so that they can gain access to all locations in a city in a reasonable time. Of course, the principles may be extended to cover other branches and individual users of the health service, but in no other case is travel time such a sensitive indicator. By examining in detail the spatial issues involved in the provision of A & E services, a fuller understanding of the accessibility problem is obtained. A key difference between A & E services and other hospital-based services, however, is that patients are not usually in a position to select the destination or hospital of their choice. In Chapter 4, where we look at the demand for hospital services, the effects of demand and access on spatial choice are examined from a more behavioural standpoint.

In this discussion, methods are presented that enable a new districting criterion to be devised, in which the objective is to locate health facilities so that no individual is more than a given journey time from a specified location. The results are thus a direct corollary of the MC-districting criterion described in Chapter 2, which in turn is related to the hexagonal patterns typifying central place theory. Like the transformation methods that were subsequently used to develop the P-criterion, related procedures are needed to develop districting patterns based on travel time. One critical difference between the two districting concepts, however, is the question of permanence. Whereas population shifts take place over a relatively long period, travel times vary on a daily basis, and are influenced by a wide variety of factors, including peak-hour travel effects, and poor weather. This means that the districts appropriate for one set of factors are not appropriate for another. For A & E services, the situation is further complicated by the changing probability of accidents occurring at different times of the day, so that administrative measures aimed at ensuring fullest use of resources are more difficult to evolve. This variability underlines the need for flexibility and good management, in particular knowing the optimum set of emergency facilities to open at particular times of the day and under different operating conditions consistent with service requirements. Illustrations of these issues and their consequent impact on the allocation of resources will thus serve to underline the nature and importance of health facility location.

To start the discussion it is necessary to establish the general geometry of movement in cities in very general terms. This will provide the basis for the analysis of districting problems which follows. After this the results are applied to accident and emergency services using data from London as the basis.

3.2   The geometry of movement in cities

The theory of movement in cities described by Angel and Hyman (1976) may be summarised as follows. If each point in a city is assigned a velocity that is independent of direction and varies continuously from one point to another, the travel time t on a path of length s between two arbitrary points A and B can be written as

$$t = \int_A^B \frac{ds}{V(r)} \qquad (3.1)$$

where $V(r)$ is a radially symmetric function describing the increase in the average speed of travel V with distance r from the city centre. The character of $V(r)$ is discussed in more detail later, but in general it assumes that central speeds are less than those at the edge of the city due to traffic congestion and the road configuration in central city areas. To find the quickest journey time between A and B, it is necessary to find the smallest value of this integral. This is a typical problem in the calculus of variations and it gives rise to a function t of the form

$$t = t(r_A, \theta_A, r_B, \theta_B, a_1, \ldots, a_n) \tag{3.2}$$

where r_A, θ_A, r_B, θ_B, are the coordinates of A and B and $a_1, \ldots, a_n$ are the parameters of $V(r)$, the velocity field.

A further definition that can be made at this stage is of an *isochrone*, which is the locus of points that can just be reached in a given time from a specified point. By considering A as the location of the facility, t as the journey time, the isochrone around the facility can be drawn (Mayhew 1981). An isochrone is the time equivalent of a hospital district delineated on the MC principle (Ch. 2), which was based on the notion of a locus of points a given *distance* from a facility, i.e. a circle of constant radius. Points inside an isochrone can access a facility in less than t; points outside must have access to a facility in times greater than t. There exist several classes of analytic forms for the equations of isochrones that are helpful and realistic in dealing with the problems with which we are concerned. These are summarised in separate mathematical notes at the end of the chapter. Unlike the simpler districting methods of Chapter 2, the resultant isochrones vary in size and shape with circles arising only in special circumstances.

3.3   Time surfaces

One convenient method of calculating the minimum journey times, quickest paths and isochrones is based on the notion of a *time surface*. This can be imagined as a portion of the physical surface of the city that has been transformed into another surface on which shortest paths (geodesics) correspond to the quickest paths in the city. Four examples of such surfaces are the plane, cylinder, cone and sphere.

It is shown in Angel and Hyman (1976), Hyman and Mayhew (1982, 1983), and in the mathematical notes at the end of this chapter, that these surfaces correspond respectively to the following four fields:

plane	$V(r) = \text{constant}$		(3.3)
cylinder	$V(r) = \omega r$		(3.4)
cone	$V(r) = \omega r^p$	$(0 < p < 1)$	(3.5)
sphere	$V(r) = \alpha r^2 + \beta$		(3.6)

In these equations ω, p, α and β denote fixed parameters related to the nature of the transport network in the city and the degree of traffic congestion. They may represent the long term, average state of the system based on road efficiency and transportation technology. For more detailed purposes (such as the monitoring of A & E services), they can be made to reflect the diurnal changes in traffic flows. With the exception of Equation (3.3), they all presume an increase in average speeds with distance from the city centre. Experience has shown that the four fields cover a wide range of possible behaviour, but that the power law in Equation (3.5) is the most flexible for studying the likely variations in the efficiency of movement in a city. This is because different combinations of the parameters ω and p enable a very wide range of speed–distance relationships to be considered covering a wide range of cities.

Figure 3.1 shows how the shortest paths on the time surface are transformed into the quickest paths for a point south of the city centre for the velocity fields described by Equations (3.3) and (3.4). For Equation (3.3), straight lines on the time surface, a plane, are simply transformed into straight lines in the city (Fig. 3.1a). The geometry of the time surface is hence Euclidean, with journey times proportional to straight line distances and yielding circular isochrones. In contrast, shortest paths on the surface of the cylinder between two non-centrally located points are transformed into curves that spiral around the city centre, trading off the increased journey distance against the reduction in travel time (Fig. 3.1b). Isochrones in this case enclose smaller areas near the city centre, increasing in size towards the periphery. Further details of these examples, including the derivations of the quickest paths, times, and isochrones, are given in the appendix of mathematical notes at the end of this chapter.

In Figure 3.1c a pattern of quickest paths is shown which is a combination of the patterns in Figure 3.1a, b. In this case, which is only one of many possible combinations, there may be a choice of quickest routes involving the decision to use public or private transport or different forms of public transport. In the southern sector AOB, the quickest paths are curved concavely with respect to the city centre as was the case in Figure 3.1b. In the remainder of the city the routes are shown as straight lines; thus, to accomplish a journey to a point, Q say, in the northern sector, the individual must first break his journey at the city centre, O, as he cannot reach Q directly. The curved lines joining AO and OB represent an indifference curve, since the appropriate journeys can be undertaken in the same journey time either by proceeding directly or by travelling first to the city centre. An improvement in peripheral travel efficiency would be represented in this diagram by an enlargement of the southern sector AOB. An increased number of journeys would be made directly, reducing the use made of the city centre as an interchange point. More locations in the suburbs could then be reached than before within a given journey time. This would enlarge the choice of hospitals for patients and reduce the locational attractiveness of the city centre. One final important

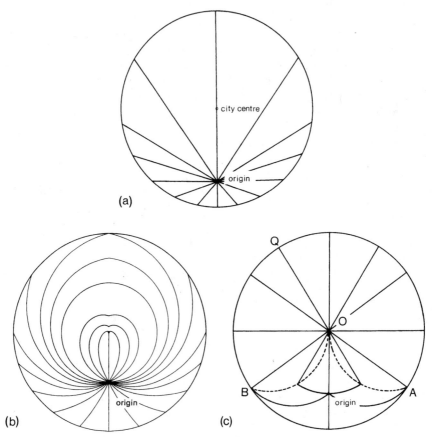

Figure 3.1 Quickest journey paths in (a) a uniform field (where $V(r)$ is constant) and (b), (c) two non–uniform velocity fields (in (b), $V(r) = \omega r$; (c) is the mixed case). In (a) shortest paths are straight lines; in (b) they curve around the city centre to avoid the worst effects of congestion. In (c) journeys proceed via the city centre if the destination is outside sector AOB, in which case journeys are straight lines. Inside sector AOB paths are again curved. The size of sector AOB depends on the relative efficiency of circumferential travel versus radial travel.

consideration will be the interchange time at the city centre in the case of public transport. If the frequency of public transport services is low, a further time penalty is incurred that is independent of the speeds of the vehicles. To conclude therefore, it would be valid to say that an increased usage of private transport should, in the long term, have a dispersing effect on hospital services. By contrast, public transport services should have a concentrating effect on their locations.

In terms of isochrones, the third example would yield a similar pattern to those for Figure 3.1b, provided that they are contained within the sector AOB. They would be small, off-centred circles in the vicinity of the city centre, increasing in radius at the edge of the city. The interesting case arises when the isochrones cut the curves of indifference because a second portion to the original isochrone is adjoined. This is a circle centred on O, which is the locus of all points that can be reached in a given time for journeys diverted through the city centre. This circle joins on the part of the isochrone previously contained in AOB to form two cusps. The lines of indifference then create two curves of cusps AO and OB. The overall effect is of two circles that are stuck non-smoothly on to one another.

A final interesting corollary of these illustrations concerns those cities in which the street patterns form a grid. For these cases, the continuity assumption employed above may be inappropriate, since travel takes place in rectilinear movements according to a Manhattan distance metric, named after the area in New York with such a road network. Without entering into detail, it is easy to show that isochrones in these cases are approximated by shapes resembling diamonds in the simplest cases. It is taken for granted that concepts developed in this chapter could be refined to deal with such cities, if appropriate modifications were made.

3.4   The perception of distance

It is instructive before proceeding with other technical issues to visualise how a variable accessibility could influence hospital choice behaviour as seen through the eyes of the patient. The easiest way to proceed is to imagine each patient as possessing a mental map of the city by which he or she judges the nearness (and hence convenience) of different hospitals. These maps are such that the usual conception of what is near and what is far, are completely at variance with how a usual map would appear. Moreover, the map would look different depending at which point in the city a patient started his or her journey and the means of transport that were available.

The following examples are illustrative of what might be entailed in the case of someone accustomed to travelling by car. In constructing them the base pattern of hospital provision, a P-districted city, is chosen, based on Figure 2.6. It will be recalled that this pattern of provision was declared efficient from the point of view of the provider but not equitable from the point of view of the patient. Figure 3.2 shows the same districting pattern, in which the distance from a point has been scaled in proportion to travel time for one particular velocity field. The reason why the patterns are different is because, in effect, space is distorted by different amounts depending on location, so that a new map has to be created for each point. The locations in question are 0, 7 and 25 km from the city centre, where each may be considered as a possible journey origin of someone seeking hospital treatment at any one of the hospitals available. The results show that the proximity to services is greatly affected. At the centre (Fig. 3.2a) the pattern is symmetrical because the ease of

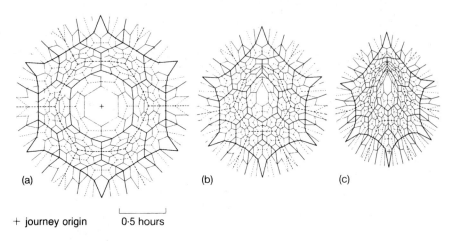

+ journey origin   0·5 hours

Figure 3.2   Theoretical maps of travel times from three points in a city: (a) 0 km, (b) 7 km, and (c) 25 km from the city centre. Distance from these points to other points on the maps is scaled to the predicted journey times. The maps show how the ease of access to other points in a city varies according to location.

travel is the same in all directions. This is not true of other locations, where the degree of distortion is dependent on local speeds in different directions. In Figure 3.2b, for instance, the location considered is 7 km south of the city centre. Because the centre represents a barrier to through-travel, points diametrically opposite are relatively distant in terms of travel time. Another key difference is the size of the lowest level central district in Figure 3.2a. This is now very large illustrating the important paradox that people living in the centres of cities and using cars may have very poor access to hospital services, in spite of their relative proximity. In Figure 3.2c, the final example, the city resembles the shape of a raindrop. Its size is further reduced because higher local speeds make distances seem shorter. Thus, although local districts are large in area, the improved travel times have a 'shrinking' effect. It is a debatable point whether car users are really as disadvantaged as these maps would indicate, because they can always switch to public transport which may offer narrow time savings in some instances depending on the time of day. Also, it remains a challenge to design similar maps for other modes of transport with a different spatial coverage of the city. Nevertheless, the conclusion from this short discussion is that variations in accessibility are associated with important implications for the locations of hospital services.

3.5   Districting cities with variable accessibility

With the above dilemma of providing a uniform access to hospital services as background, the discussion now turns to the issue of deriving a pattern of hospital districts so that no one is more than a fixed standard of travel time away. As in Chapter 2, it is convenient to begin the analysis by considering a radially symmetric city in which (i) hospitals are free to locate anywhere; (ii) there exists a time surface; and (iii) there are no barriers, such as rivers, to travel. (This final restriction will later be relaxed.)

To cover the time surface with the minimum number of hospital facilities, it is necessary that each serves the maximum possible area on the time surface, thereby minimising the degree of overlap between catchment areas defined by the time standard. Assuming speeds are uniform and proportional to distance, it was shown in Chapter 2 that the appropriate, optimum pattern of locations was described by a tessellation of regular hexagons, since regular hexagons have the maximum ratio of area to radius (see also Sec. A2, appendix).

When speeds are not uniform, it would be expected that distortions in this pattern would take place locally, with the area of each hexagon increasing with distance from the city centre. From the definition of the time surface (Sec. 3.3), however, equal-sized hexagons on the time surface would correspond to the unequal hexagonal districts in the physical plane of the city. The problem, therefore, is to find a suitable transformation from the time surface to the plane that correctly portrays this distortion.

3.6   Tessellating the cylinder, cone and sphere

If the velocity field is $V = \omega r$, the time surface is a cylinder, and the city can be optimally covered with a constant number of facilities around each ring of constant radius (see Sec. A2). If the velocity field is $V = \omega r^p$, the time surface is a cone that can be exactly tessellated only for restricted values of p. There are 13 possible cases: $0, \frac{1}{6}, \frac{1}{3}, \frac{1}{2}, \frac{2}{3}, \frac{5}{6}, 1, \frac{7}{6}, \frac{4}{3}, \frac{3}{2}, \frac{5}{3}, \frac{11}{6}, 2$. The cases 0 and 1 correspond to the plane and cylinder already discussed while those where $1 < p \leqslant 2$ correspond to inverses of the cases $0 \leqslant p < 1$ (see Sec. A2). Unless p is less than unity, the number of facilities required diverges as the city centre is approached. For the cases where $p < 1$ the conical tessellations can be constructed by cutting sectors out of a plane hexagonal tessellation and gluing the cut edges back together.

If the velocity field is $V = \alpha r^2 + \beta$, the time surface is a sphere. In this instance it is impossible, however, to fit a hexagonal tessellation exactly on to the entire surface (see Sec. A2). In practice only a portion of the sphere would need to be covered, so this is not a serious restriction. However, it is of theoretical interest to investigate tessellations that *are* efficient covers of the entire sphere. The two best examples are the dodecahedron and the truncated icosahedron. The former is a regular Platonic solid, which contains 12 pentagonal faces; the latter is a solid with 12 pentagonal faces and 20 hexagonal faces (often used in the construction of soccer balls). However, these could only form solutions to problems of optimum facility coverage if the velocity field

Figure 3.3 The basic tessellated surfaces including the plane, five cones, a cylinder and two spheres.
Source: Hyman and Mayhew 1983.

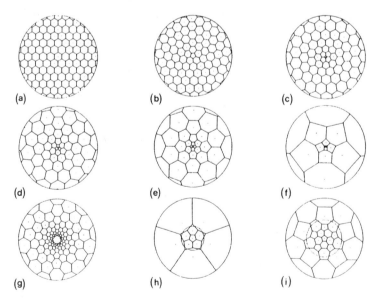

Figure 3.4 The tessellations transformed to the physical surface of the city. The time taken to travel from the vertex of any boundary to the contained hospital facility by the shortest path is the same for a given tessellation.
Source: Hyman and Mayhew 1983.

Table 3.1   Key to the tessellations shown in Figure 3.4.

Figure number	Velocity field (km h^{-1})	Journey time (h)	Surface
3.4a	17.0	0.229	plane
3.4b	27.0$r^{1/6}$	0.096	cone
3.4c	15.0$r^{1/3}$	0.119	cone
3.4d	8.0$r^{1/2}$	0.164	cone
3.4e	4.5$r^{2/3}$	0.242	cone
3.4f	2.5$r^{5/6}$	0.475	cone
3.4g	1.5r	0.155	cylinder
3.4h	0.77r^2 + 5.0	0.167	sphere
3.4i	0.31r^2 + 5.0	0.169	sphere

$V = \alpha r^2 + \beta$ and the time standard happened to be scaled in suitable proportions (see Sec. A2).

Figure 3.3 is a photograph of all the tessellated surfaces described so far, the cones covering the cases $0 < p < 1$. For clarity, only a one-level hierarchy is considered. Figure 3.4, by contrast, shows the same tessellations after they have been transformed to the physical surface of the city. (For details of the transformations, see Sec. A1.) The scale of the diagrams depends on the form and parameters of the velocity field, and the journey time t. Table 3.1 gives the values chosen for these examples. There are a number of points to note concerning the results:

(1)   To simplify the construction, the boundaries of the districts are shown as straight lines. While this is an adequate approximation at this scale of inquiry, they should more correctly be shown as curves.

(2)   In the cylindrical transformation ($V = \omega r$), the number of facilities is constant in each ring, so that the number of facilities becomes infinite at the city centre. Since the demand for services is finite it would be impossible to maintain such a pattern in this locality.

(3)   For the conical transformations ($V = \omega r^p$, including the plane as a special case) the shape of the central facility catchment polygon depends on the value of p. The possible shapes are

p	central catchment polygon
0	hexagon
1/6	pentagon
1/3	square
1/2	triangle
2/3	biangle (◗)
5/6	monoangle (Ö)

The number of facilities packed around the city centre increases in successive rings. It varies with p according to $6n(1 - p)$, where n is the ring ($n = 1, 2, \ldots$) and where the first ring is the one touching the central polygon.

(4) For the two examples shown of the spherical transformation ($V = \alpha r^2 + \beta$), the tessellations are either all pentagons or a mix of pentagons and hexagons. In both cases, the facility antipodal to the central facility cannot be shown as it is located at infinity. When the tessellations consist of mixed figures, as in the case of the 'soccer ball', time standards are not uniformly achieved $-t_{pentagon}$ being less than $t_{hexagon}$. The dotted circle is the equator.

(5) If these districting patterns are used to develop the hierarchical patterns described in Chapter 2, similarly shaped central catchment polygons will be obtained at successive levels. For the spherical transformations only, the next level up from the truncated icosahedron (Fig. 3.4i) would be the dodecahedron (Fig. 3.4h), otherwise the number of edges per district is unaffected.

These illustrations show that it is possible to construct optimal patterns for the location of hospital facilities to meet a consistent travel time standard when travel speeds vary across a city. The most flexible class of velocity field was seen to be particular types of power law variations of velocity with distance. These are associated with conical time surfaces and, unlike the examples given for the sphere, are not subject to scale restrictions. Furthermore, the conical time surfaces can be exactly covered with hexagons at all locations apart from the city centre.

It is also possible to develop other types of velocity fields for cases where one of those described above does not adequately fit the pattern of travel in an urban region. One technique, for instance, is based on gluing together segments of the time surfaces to produce more realistic profiles of average speeds in given situations. For example, a problem with the power law (Eqn 3.5) is that it predicts a zero speed at the city centre, with indefinitely increasing speeds at large distances. Although not a serious restriction, a zero speed seems unlikely even in the most congested of cities, whilst large speeds are restricted by law and by the physical capabilities of road vehicles. Gluing is a way of overcoming these difficulties through the appropriate choice of segments from standard fields to provide a more realistic model.

3.7 Barriers to travel

The districting patterns described depend on an analogy between the network and a surface on which velocities change smoothly between points. Such generalisations are often adequate for a broad understanding of the basic forces influencing the locations of health care facilities. At a more detailed scale of analysis, obstacles to travel arise causing discontinuities in the velocity field and these should also be taken into account. For instance, a hospital located by a river which has only thinly spaced crossing points has a portion of its catchment area effectively truncated. Similarly, a hospital located next to a large park, reservoir or industrial estate with restricted access is locationally less efficient than one with unrestricted access. Clearly, the geographical distribution and scale of different barriers to travel will vary from city to city as will the extent of their hinderance. Although generalisations are therefore impossible to make, it is possible to reduce the many different types of barriers to one of two basic classes: (a) line barriers, such as rivers and railways, where crossing is restricted to selected bridging points, and (b) open expanses, which are closed to traffic and have to be circumnavigated.

Line barrier
Figure 3.5 shows how a line barrier causes a discontinuity in the pattern of isochrones radiating from point O. As is seen, the barrier disrupts the isochrones, except at A and B (two bridges), which become foci for two new sets of isochrones on the opposite side. This forces the new sets to contain

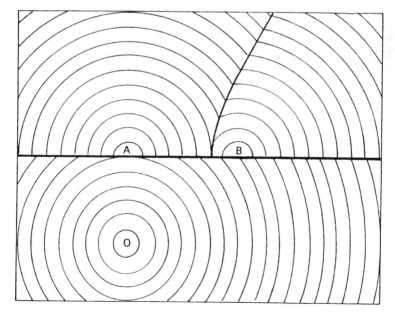

Figure 3.5 A line barrier, such as a river, showing isochrones radiating from O, a journey origin. The thick line is the river, and A and B are two crossing points. The line joining the cusps on the opposite bank divides quickest routes to destinations on the left via A or on the right via B.
Source: Hyman and Mayhew 1983.

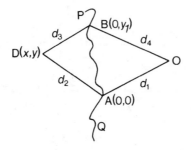

Figure 3.6   Two routes across a river PQ between O and D.
Source: Hyman and Mayhew 1983.

cusps at the points of contact. The curve joining the cusps is the line of indifference, points on which can be reached in the same time via bridge A or B. To show this, briefly consider Figure 3.6. PQ is the river, O and D are the origin and destination, and A and B are the bridge locations as before. If the speed of travel is locally constant, then travel from O to D takes the same time via either bridge. That is when

$$d_1 + d_2 = d_3 + d_4 \tag{3.7}$$

Let $d_4 - d_1 = \varepsilon$, then

$$\varepsilon = d_2 - d_3 \tag{3.8}$$

$$= \sqrt{\{x^2 + y^2\}} - \sqrt{\{x^2 + (y - y_1)^2\}} \tag{3.9}$$

Thus

$$x^2 + y^2 = \varepsilon^2 + x^2 + (y - y_1)^2 + 2\varepsilon\sqrt{\{x^2 + (y - y_1)^2\}} \tag{3.10}$$

Simplifying and ordering the terms,

$$4\varepsilon^2 x^2 + 4(\varepsilon^2 - y_1^2)y^2 + 4y_1(y_1^2 - \varepsilon^2)y = (\varepsilon^2 - y_1^2)2 \tag{3.11}$$

Equation (3.11) is a hyperbola when $|y_1| > \varepsilon$. This inequality will be satisfied unless D lies on either extension of the line segment AB joining the bridges. In this case $y_1 = \varepsilon$, and the equation becomes $x = 0$; $y \leqslant 0$, or $y \geqslant y_1$, which describes the two extensions of the line segment AB. Points to the left of the hyperbola are reached quickest via bridge A and points to the right via bridge B.

Open expanses

The approach to dealing with open expanses is broadly similar to that for line barriers. In Figure 3.7 an open expanse is shown represented by the triangle ABC. Each vertex of the triangle is the starting point for a change in the pattern of isochrones for the areas that lie in the shadow of the journey origin at O. The isochrones intersect on the far side of the barrier and a similar curve of cusps to the bridge example is generated.

Clearly, which barriers to include or exclude is entirely a matter of how accurate the estimates of travel time need to be. To include every single barrier would be hopelessly complex, but to ignore major barriers would lead to poor results. Hence, the matter is essentially empirical and should be judged in accordance with the intended use of the analysis. In the next section the theory developed in this chapter is applied to A & E services. The analysis builds on the location of services currently provided in London but no particular account is taken of certain significant barriers in this city, such as the river Thames. In this case, most of the corrections needed could be made by local scale modifications to the isochrone patterns.

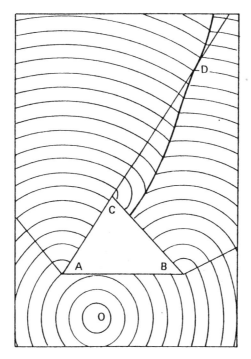

Figure 3.7   An open expanse ABC as an obstacle to travel. The journey origin is O. A curve of cusps through D divides quickest routes as in Figure 3.5.
   Source: Hyman and Mayhew 1983.

3.8   Emergency medical services: a practical illustration

The next task is to place the above concepts in a practical perspective, and for this the example of accident and emergency (A & E) services is used. It goes without saying that in actual cities the neat and exact location patterns presented earlier will never occur. On the other hand, the need to provide an even service standard, based on uniform access in all areas, will lead to a pattern of provision which may provide a reasonable approximation. As an illustrative evaluation of some actual patterns, isochrones generated using the theory are drawn around selected treatment centres based in London for different assumed operating conditions, where the objective is to show whether or not there are any geographical gaps in provision. The discussion then moves to the resource implications (in terms of case-load and other factors) during the day and night operation of A & E services under different assumed service standards. For these purposes an improved estimate of the London velocity field is used. This leads to a number of important implications which are then considered in the light of the theory.

Isochrone maps of emergency treatment facilities

Figure 3.8 shows instances of *actual* arrangements of emergency treatment centres in London within the Greater London Council area. Similar maps could also have been drawn for ambulance depots, although the same locational principles discussed in these sections would apply. Around each centre a 10 min isochrone has been drawn to indicate the predicted operating range of emergency vehicles. Locations inside isochrones are within 10 min of a centre; locations outside isochrones take more than 10 min to get to a centre. The velocity field which is assumed in these illustrations is given by

$$V = \omega r^p \qquad (0 < p < 1) \tag{3.5}$$

where V is the average velocity of ambulances in kilometres per hour at a distance r from the city centre. The parameters ω and p are altered in each map to create the hypothesised differences in the operating conditions presumed to result from varying traffic congestion. Near the city centre the isochrones enclose smaller areas than in peripheral locations because in this area traffic congestion reduces travel speeds and hence the operational effectiveness of emergency vehicles. The mathematical derivation of the equations for the isochrones shown is given in Section A3 at the end of the chapter. Normally, the isochrones are off-centred circles, flattened on the side nearest the city centre. Near the city centre itself, however, they can distort to other shapes due to the rapid decline in central area travel efficiency. The maps show that, as the operating conditions vary, different numbers of facilities need to be open to provide an adequate coverage of the region at the 10 min standard. In the examples, locations are selected if they do not lie within the isochrone of a

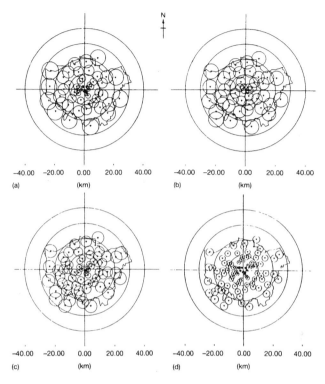

-40.00 -20.00 0.00 20.00 40.00   -40.00 -20.00 0.00 20.00 40.00
(a)   (km)   (b)   (km)

-40.00 -20.00 0.00 20.00 40.00   -40.00 -20.00 0.00 20.00 40.00
(c)   (km)   (d)   (km)

Figure 3.8 Ten-minute (0.167 h) isochrones around selected medical centres in London for four different parameter sets. The number of centres needed to cover the urban area adequately at a given time standard varies according to traffic conditions. In all cases the areas enclosed by isochrones near the city centre are smaller than those in the suburbs. (Parameter values: (a) 53 facilities, $\omega = 3$, $p = 0.75$; (b) 44 facilities, $\omega = 10$, $p = 0.33$; (c) 66 facilities, $\omega = 5$, $p = 0.5$; (d) 90 facilities, $\omega = 3$, $p = 0.5$).
Source: Hyman and Mayhew 1983.

neighbouring facility. This indicates the approximate number required, although not which ones exactly. In all the examples there are gaps between the isochrones representing local inefficiencies in the distribution of facilities and particularly large gaps show where new services are advisable, although in terms of the optimum, some inefficiency is always to be expected because of the highly variable nature of traffic conditions. Also some gaps in these particular examples are explained by the presence of parks, reservoirs, etc. Nevertheless, if traffic conditions deteriorate sharply as is implied by Figure 3.8d, then access standards may be impossible to maintain without the opening of many new treatment centres. Finally, it is noteworthy that in each case there is a tendency for centres to overcrowd areas near the city centre. In accessibility terms, this would indicate that there is considerable scope for a rationalisation of the facilities provided.

Further estimates of the daily ambulance velocity field

Figure 3.8b, c are closest to the operating conditions that are actually experienced in this particular city. Following more detailed work (Hyman & Mayhew 1983) using ambulance travel times, it was established that a more appropriate characterisation of average ambulance travel speeds in London could be accurately described by the following two fields, one operating during the day and evening and the other at night. They are:

Day and evening

$$
\begin{array}{llr}
V(r) = 8.8 & (0 \leqslant r \leqslant 4) & (3.12) \\
V(r) = 4.4r^{1/2} & (4 < r \leqslant 15) & (3.13) \\
V(r) = 1.13r & (15 < r \leqslant 25) & (3.14) \\
V(r) = 28 & (r \geqslant 25) & (3.15)
\end{array}
$$

where V is expressed in kilometres per hour and r in kilometres.

Night

$$
\begin{array}{llr}
V(r) = 17 & (0 \leqslant r \leqslant 15) & (3.16) \\
V(r) = 1.13r & (15 < r \leqslant 25) & (3.17) \\
V(r) = 28 & (r \geqslant 25) & (3.18)
\end{array}
$$

These two cases are examples of the 'gluing' principle mentioned above in Section 3.6. Thus, the day-time field has a constant velocity out to 4 km from the centre; thereafter it varies, first according to the square root of distance and then in direct proportion to distance. At 25 km it is constant at 28 km h^{-1}. At night the central velocity is higher than for the day; moreover, the constant section reaches out to 15 km. Beyond 15 km the night-time field is the same as the day-time field. The appropriate time surface for this field consists during the day of a flat section, a conic section, a cylindrical section, and then another flat section, whereas at night, the conic section is missing. These fields are now used to establish various characteristics of the A & E services that cover the city. The first task, however, is to define the key aspects of demand and supply. This is carried out using a similar, generalized approach to that employed in Section 2.10. The following definitions are established:

(1) $D(r, T)$ is the density of population distance r from the centre, time T of the day
(2) $\lambda(r, T)$ is the daily probability of generating an accident or emergency per capita
(3) $C(r, T)$ is the expected number of accidents and emergencies per unit area.

It follows that

$$
C(r, T) = \lambda(r, T) D(r, T) \tag{3.19}
$$

(4) $A(r)$ is the area of a 1 km wide ring $(=2\pi r \times 1)$

(5) $N(r, T \mid t)$ is the smoothed number of facilities in a 1 km wide ring given t, the required time standard for accessing facilities.

(6) $\sigma(r, T \mid t)$ is the catchment area of an optimally located treatment facility.

An approximate estimate for σ is given by

$$\sigma(r, T \mid t) = V(r, t)^2 3\sqrt{3}t^2/2 \qquad (3.20)$$

where $V(r, T)$ is the local velocity; the remaining terms give the size of a hexagonal catchment area on the time surface.

(7) $M(r, T \mid t)$ is the average daily number of patients per facility by time of day for a given time standard.

Definitions 1–3 describe the demand side of the A & E services over the area of the city and at different times of the day, whereas definitions 5 and 6 are more concerned with the supply side. Definition 7, the expected case-load at different facilities, is the anticipated outcome of different combinations of supply and demand for a given time standard. Lastly, from Definitions 4 and 6, it is also seen that

$$N(r, T \mid t) = \frac{A(r)}{\sigma(r, T \mid t)} \qquad (3.21)$$

Hence, it further follows, from Definitions 3, 4 and 5, that

$$M(r, T \mid t) = \frac{A(r) C(r, T)}{N(r, T \mid t)} \qquad (3.22)$$

The number of A & E facilities needed for different time standards
The specification of an access time standard t for an optimally located set of A & E facilities determines the number of locations needed under different operating conditions. To show this, it is noted that the area of a regular hexagon drawn on the time surface of the London velocity field is given by

$$v = \frac{3\sqrt{3}}{2}t^2 \qquad (3.23)$$

where v is expressed in minutes squared. The estimated optimum number of treatment centres is obtained by dividing the total area of the time surface corresponding to the city by the area of the hexagon for which t is the access time standard. The result shows that the required amount must be inversely proportional to the square of the time standard.

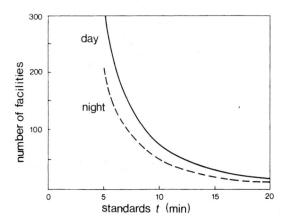

Figure 3.9 Variations in the number of treatment facilities with operating time standards, t, by day and night. The curves show the minimum number of facilities necessary to cover the London area completely at a given time standard.
Source: Hyman and Mayhew 1983.

Figure 3.9 illustrates this result for the London case based on the two velocity fields described above, one operating at night and the other during the day and evening. On the vertical axis are the numbers of facilities; on the horizontal axis is the time standard. Two curves are shown because at night congestion, and hence travel times, are greatly reduced, so that on average fewer facilities are needed. The equations of the two curves are $N = 7700/t^2$ (day-time) and $N = 5000/t^2$ (night-time), so that for a 10 min. standard, say, 77 facilities are needed during the day and 50 at night. From the two curves, it is also plain that considerable sensitivity exists to changes in operating conditions. For example, if traffic conditions were to worsen so that travel speeds were reduced by 10 per cent and a 10 min. time standard was in operation, then there would have to be a two min. delay in access times or there would need to be a 20 per cent increase in the minimum number of facilities.

Figure 3.10 shows the same result in a different way. It contains graphs of the minimum number of A & E facilities that need to be available at different distances from the centre of London for varying access time standards of 9, 10 and 11 min. Equation (3.21) has been used in the calculation of the graphs which show for the day-time case a steady increase at first in the facilities needed, then a levelling off, followed by a decline and another increase. At night far fewer facilities are needed in most areas because there is less traffic congestion. The discontinuities in the graphs correspond to the break-points in the velocity fields at the distances shown. Thus, in the level middle section the underlying average ambulance speeds are increasing in proportion to the radius, so that only a constant number of facilities on each ring around the

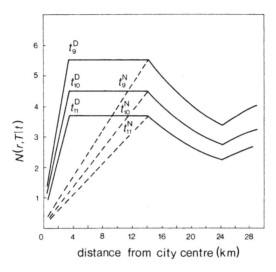

Figure 3.10 The variation in the minimum number of facilities N $(r, T | t)$ needed at different distances from the city centre by night (N) and day (D) given time standards, t, of 9, 10 and 11 min.
Source: Hyman and Mayhew 1983.

centre is required. As in Chapter 2, this exercise has shown, therefore, the importance of access as a determinant in the provision of an effective service. At the same time, however, it has raised other problems regarding the resources that need to be on standby in different locations, and these are now analysed in more detail.

The nature of demand for A & E services

The usage for A & E services is dominated by chance, so that in any one hour it may be impossible to predict how may patients will need attention. Over longer periods, however, there is often sufficient regularity in the pattern of demand to permit a certain degree of planning. The distribution of risks of involving accidents varies according to population density and the locations of different risk groups (e.g. commuters, shoppers, the elderly, school–children). For this reason the pattern at night will differ from that during the day and at weekends. Data from the London area in 1979, for example, show that the risk of accidents occurring increases steadily through the morning reaching a peak around 1200 hours, and then declines slowly, whereas at night demand drops dramatically. Estimates of the incidence of emergency calls in different areas of the city can also be conveniently expressed in relation to distances from the centre of the city in the same way as the population density (see also Eqn 3.19). This relationship which shows a negative expotential decline in activity from the city centre, enables a simple derivation of the expected case load at different emergency centres by day or night for different access time standards. In the

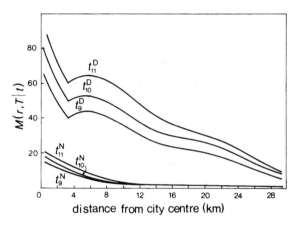

Figure 3.11   The expected case-load M at a facility optimally located distance r from the city centre during time period T at given time standards t of 9, 10 and 11 minutes (where D is day and N is night).

particular case of London, the number of night-time emergency calls (02.30–06.30 hours) is on average less than one third of the activity during the remainder of day for a typical Monday to Thursday period. On the other hand, it has been shown that access in central areas of the city is more efficient at night because there is far less traffic so that fewer facilities are needed. In terms of Equation 3.22, the above considerations produced the set of case-load variations shown in Figure 3.11, which is a plot of the expected case-load on distance from the city centre for an optimally located treatment centre. As is seen, the case-load is always highest near the centre, with a second peak, for daytime only, at about 6 km. Not unexpectedly a smaller time standard is also associated with a smaller case-load. At night emergency calls are less frequent, and so the case-load drops dramatically despite there being fewer facilities, particularly towards the edge of the city. In the latter instance, the question arises of whether it is economically justifiable to make certain treatment centre facilities available at all times. However, it has to be borne in mind that if some were to close, then access standards in the areas affected would inevitably fall. Interestingly, this represents precisely the same disadvantage noted during the discussion in Chapter 2 of the relative merits of the MC-districting criterion. In the latter case, however, the disadvantage would be greater, since the efficiency of access is presumed not to vary between different areas of the city.

   In summary, then, this brief illustrative application has shown the flexibility of the districting approach in dealing with one particular location problem. Plainly, there is much further experimentation that could be done using the concepts described. Perhaps the key factor to emerge, however, is the critical nature of the accessibility factor in shaping particular health services and this is a theme which is addressed in a more general context in the next chapter.

3.9 Conclusions

The concept of accessibility was redefined in terms of travel time in order to examine the possible impact on health facility locations of variations in travel efficiency. To avoid the complexities involved in the measurement and interpretation of travel time, the concept of a velocity field was employed to ilustrate the general geometry of movement in cities, including an indication of quickest routes and how to evaluate isochrones. The use of journey time was seen to generate some surprising effects depending on the mode of travel and the location of the traveller in relation to his or her destination. Variations in travel efficiency, for example, were seen to cause possible differences in the perceptions of nearness to hospital that were found to have some counter-intuitive implications.

To overcome some of the effects discussed, a form of MC-districting was proposed based on a pattern of locations in which no part of the city was more than a fixed standard of travel time away. It was shown that tessellations drawn on a time surface and then transformed to the physical surface of the city were able to achieve this objective using the minimum possible number of facilities. It was suggested that the criterion of uniform access time was also valid for locating A & E services (ambulance depots and treatment centres) and some hypothetical tessellations were given as illustrations.

The concepts and methods described in the early sections of the chapter were then applied to the problem of providing A & E services in London. It was shown that, based on reasonable assumptions regarding vehicle speeds, the central areas of the city were substantially over-provided with emergency treatment facilities at an access standard of 10 minutes. The scope for rationalising facilities in suburban areas appeared to be less, however, and in some places geographical gaps in provision were identified. In terms of allocating resources, the results indicated that fewer resources needed to be deployed at each centre in suburban areas, as the chances of an emergency arising were less than in the central area.

Appendix to Chapter 3: mathematical notes

The mathematical notes presented here are based partly on a book by Angel and Hyman (1976), entitled *Urban fields*, on the paper by Hyman and Mayhew (1982) 'On the geometry of emergency service medical provision in cities', and on 'Automated isochrones and the locations of emergency medical services in cities' by Mayhew (1981).

A1   Radially symmetric fields

We are given a polar coordinate system (r, θ), for the urban plane, on which a velocity field $V(r)$, has been defined, and a cylindrical coordinate system (ρ, z, ϕ), for the space in which the time surface is located.

By Theorem 3.1 in Angel and Hyman (1976, p. 45), the following changes of variable,

$$\phi = \theta \qquad (A1.1)$$

$$\rho = r/V \qquad (A1.2)$$

and

$$z = \int \frac{1}{V^2} \left[2rV \frac{dV}{dr} - r^2 \left(\frac{dV}{dr} \right)^2 \right]^{1/2} dr + C \qquad (A1.3)$$

define a transformation

$$T_V : (r, \theta) \rightarrow (\rho, z, \phi)$$

This transformation maps travel time on any path P on the urban plane onto the length of the image of that path $T_V(P)$.

Corollary 1.1 *The time surface of the velocity field $V(r) = \omega r$, where ω is a constant, is a cylinder.*

PROOF From (A1.2) and (A1.3),

$$\rho = 1/\omega \qquad (A1.4)$$

and

$$z = \int \frac{dr}{\omega r} + C = \frac{1}{\omega} \ln \left(\frac{r}{r_0} \right) \qquad (A1.5)$$

where r_0 is the radius corresponding to $z = 0$. ρ is independent of z, and the time surface is thus a cylinder of radius $1/\omega$ that extends over the complete z axis, from $z = -\infty$ (corresponding to $r = 0$) to $z = +\infty$ ($r = \infty$).

Corollary 1.2 *Journey time t in the field $V(r) = \omega r$ is given by*

$$t_{12} = \frac{1}{\omega} \left[\ln \left(\frac{r_2}{r_1} \right)^2 + (\theta_2 - \theta_1)^2 \right]^{1/2} \qquad (A1.6)$$

PROOF Open a cylinder along a generator and lie it flat on the ground. From the definition of a time surface, the minimum travel time between two points is the length of the straight line connecting them. Thus,

$$t_{12} = \left[(z_2 - z_1)^2 + \left(\frac{\phi_2 - \phi_1}{\omega} \right)^2 \right]^{1/2} \tag{A1.7}$$

where $(\phi_1/\omega, z_1)$ and $(\phi_1/\omega, z_2)$ are the coordinates of the points on the cylinder. Substituting (A1.1) and (A1.5) in (A1.7) and simplifying, we obtain

$$t_{12} = \frac{1}{\omega} \left[\ln \left(\frac{r_2}{r_1} \right)^2 + (\theta_2 - \theta_1)^2 \right]^{1/2} \tag{A1.6}$$

the required result.

Corollary 1.3   *An isochrone in the field* $V(r) = \omega r$ *is given by*

$$r_2 = r_1 \exp[t^2\omega^2 - (\theta_2 - \theta_1)^2]^{1/2} \tag{A1.8}$$

PROOF   To obtain the isochrone, simply rearrange (A1.6) so that r_2 becomes the subject. By letting $(r_1\theta_1)$ be the location of the facility and t the time standard, and allowing θ_2 to vary, the desired isochrone may be plotted.

Corollary 1.4   *The equation of the quickest path in the field* $V(r) = \omega r$ *is given by*

$$\ln r = m\theta + C \tag{A1.9}$$

where m and C are constants.

PROOF   The equation of a straight line between two points $(\phi_1/\omega, z_1)$ and $(\phi_2/\omega, z_2)$ on an opened cylinder is

$$z = A + \frac{(z_2 - z_1)}{(\phi_2 - \phi_1)} \phi \tag{A1.10}$$

where A is a constant. From (A1.1) and (A1.5)

$$\ln \left(\frac{r}{r_0} \right) = A\omega + \frac{\theta}{(\theta_2 - \theta_1)} \ln \left(\frac{r_2}{r_1} \right) \tag{A1.11}$$

Letting

$$A\omega + \ln r_0 = C \tag{A1.12}$$

and

$$\frac{1}{(\theta_2 - \theta_1)} \ln \left(\frac{r_2}{r_1}\right) = m \qquad (A1.13)$$

two constants depending on the origin (r_1, θ_1) and destination (r_2, θ_2), Equation (A1.11) becomes

$$\ln r = m\theta + C \qquad (A1.9)$$

Equation (A1.9) describes a logarithmic spiral that radiates outwards from a specified origin. A diagram showing quickest paths in r, θ coordinates is given in Figure 3.1b.

Corollary 1.5 *The time surface of the velocity field* $V(r) = \omega r^p$ $(0 < p < 1)$, *where* ω *is a constant, is a cone.*

PROOF   From Equation (A1.2)

$$\rho = \frac{r}{V} = \frac{r^{1-p}}{\omega} \qquad (A1.14)$$

By Equation (A1.3), letting $C = 0$,

$$z = \frac{r^{1-p}}{\omega(1-p)} (2p - p^2)^{1/2} = \frac{\rho}{1-p} (2p - p^2)^{1/2} \qquad (A1.15)$$

Thus, z is proportional to ρ, and the time surface is a cone.

Corollary 1.6 *Journey time in the field* $V(r) = \omega r^p$ *is given by*

$$t_{12} = \frac{1}{\omega(1-p)} \{r_1^{2-2p} + r_2^{2-2p} - 2r_1^{1-p} r_2^{1-p} \cos[(1-p)\theta_{12}]\}^{1/2} \quad (A1.16)$$

PROOF   As with the cylinder, travel time is evaluated by measuring distances on the conic time surface. The distance R from the apex to any point on the cone is, from Equation (A1.14) and (A1.15), and using Pythagoras' theorem,

$$R = (\rho^2 + z^2)^{1/2} = \frac{\rho}{1-p} \qquad (A1.17)$$

The cone can be opened along a generator to obtain a pie-shaped area. The angle ϕ is transformed into the angle α between two generators, such that

$$\alpha = \phi(1-p) \qquad (A1.18)$$

(see Angel & Hyman 1976, p. 48). The time between two points (R_1, α_1) and (R_2, α_2) can be evaluated by the cosine rule. Letting α_{12} be the angular difference of the two points, then

$$t^2(R_1, R_2, \alpha_{12}) = R_1^2 + R_2^2 - 2R_1 R_2 \cos \alpha_{12} \qquad (A1.19)$$

Substituting for R_1, R_2 and α_{12} using Equations (A1.14), (A1.17) and (A1.18), the result in Equation (A1.16) is obtained.

Corollary 1.7 *An isochrone in the field $V(r) = \omega r^p$ is given by*

$$r_2 = \{m \cos[(1 - p)\,\theta_{12}]$$
$$\pm \sqrt{(\omega^2 t^2(1 - p)^2 - m^2 \sin^2[(1 - p)\theta_{12}])}\}^{1/(1-p)} \qquad (A1.20)$$

where

$$m = r_1^{1-p} \quad \text{and} \quad \theta_{12} = \theta_2 - \theta_1$$

PROOF   Equation (A1.16) may be rearranged into the standard form

$$0 = ax^2 + bx + c \qquad (A1.21)$$

where

$$a = 1$$

$$b = -2m \cos[(1 - p)\,\theta_{12}]$$

$$c = m^2 - \omega^2 t^2(1 - p)^2$$

Now

$$r_2^{1-p} = x = \frac{-b \pm \sqrt{(b^2 - 4ac)}}{2a} \qquad (A1.22)$$

Substituting for a, b and c in (A1.21) and simplifying gives (A1.20), the required result.

Corollary 1.8 *The time surface of the velocity field $V = \alpha r^2 + \beta$, where α and β are positive constants, is a sphere of radius $1/\omega$ that satisfies*

$$\left(z - \frac{1}{\omega}\right)^2 + \rho^2 = \frac{1}{\omega^2} \qquad (A1.23)$$

where

$$\omega^2 = 4\alpha\beta \qquad (A1.24)$$

PROOF   See Angel and Hyman (1976), p. 49.

A2   *Some properties of tessellations on surfaces*
There are several properties of the surfaces used in this chapter that are
essential to the arguments put forward. The properties considered below
discuss the class of admissible, optimal tessellations, restrictions on tessellating
cylinders and spheres, and admissible parameter values for exactly tessellating
cones.

Optimal tessellations   The area of any polygon drawn inside a circle so that the
vertices touch the circumference is maximised when all the sides of the
polygon are equal. Furthermore, this area increases with n, the number of
sides. We are looking, therefore, for a regular polygon that has the maximum
number of sides with which to fill a plane without any overlap. The number of
polygons needed will then be a minimum for any given radius. To determine
the plane tessellations we let k be the number of regular n-gons meeting at a
point. The angle made by each n-gon at the join is then $2\pi/k$. The internal
angle of a regular n-gon equals $\pi - (2\pi/n)$. Equating, therefore

$$\frac{2\pi}{k} = \pi - \frac{2\pi}{n} \qquad (A2.1)$$

we get

$$n = \frac{2k}{(k-2)} \qquad (A2.2)$$

The only values of k that give integer values of n are 3, 4 and 6, as is easily
verified. These correspond to a hexagon, square and triangle, respectively.
Because a hexagon has the most sides, it is, by the first argument, the required
regular figure.

Tessellating the cylinder   There are two possible orientations of a pattern of
regular hexagons that fit exactly around the circumference of a cylinder. If t is
the radius of each hexagon, then under one orientation the pattern repeats itself
in intervals of $\sqrt{3}t$; in the opposite orientation the interval of repetition is $3t$. In
either case the circumference of the cylinder is $2\pi/\omega$. An exact fit for the

tessellation will occur when this circumference equals an integer number of intervals of repetition. We therefore need to satisfy either the condition

$$2\pi/\omega\sqrt{3t} \text{ is an integer}$$

or the condition

$$2\pi/\omega 3t \text{ is an integer}$$

according to the orientation selected.

When transformed back to the city the circumferences of the cylinder correspond to radii, so that the number of facilities required is the same on each ring around the city centre.

Tessellations on the cone   The total number of *p*-cones that admit exact hexagonal tessellations is 13, namely

$$p = 0, \frac{1}{6}, \frac{1}{3}, \frac{1}{2}, \frac{2}{3}, \frac{5}{6}, 1, 2, \frac{11}{6}, \frac{5}{3}, \frac{3}{2}, \frac{4}{3}, \frac{7}{6}$$

The maps on the urban plane for the field $V = \omega r^p$ are the *inverses* of maps for the field $V = \omega^* r^{2-p}$. To show this, consider two maps from the same cone. From Equation (A1.2)

$$r = (\omega\rho)^{1/(1-p)} \qquad (p \neq 1) \tag{A2.3}$$

$$r^* = (\omega^*\rho)^{1/(1-q)} \qquad (q \neq 1) \tag{A2.4}$$

Then

$$rr^* = \omega^{1/(1-p)}\omega^{*1/(1-q)}\rho^{2-(p+q)/(1-p)(1-q)} \tag{A2.5}$$

If $p + q = 2$, we have

$$rr^* = \left(\frac{\omega}{\omega^*}\right)^{1/(1-p)} \tag{A2.6}$$

Hence

$$r^* = \left(\frac{\omega}{\omega^*}\right)^{1/(1-p)} r^{-1} \tag{A2.7}$$

We need only to construct the seven basic maps, for $0 \leqslant p \leqslant 1$, and we can get the other six from the last equation. The cylinder, $p = 1$, is a special case and is self-inverse.

Tessellating the sphere with pentagons and hexagons   The sum of the angles meeting at a point cannot exceed 2π radians. The angle of a regular pentagon is $3\pi/5$. Thus, no more than three faces can meet at a vertex. We therefore deduce the following theorem.

Theorem 2.1   *Any tessellation on the surface of a sphere using only pentagons and hexagons requires at least 12 pentagons.*

PROOF   Let p denote the number of pentagons and h the number of hexagons. For a tessellation on the sphere, let f be the number of faces, r the number of vertices and e the number of edges. The total number of faces is given by

$$f = h + p \tag{A2.8}$$

Each hexagon has six edges, each pentagon five edges and each edge is shared by two faces, so

$$e = (6h + 5p)/2 \tag{A2.9}$$

Each hexagon has six vertices and each pentagon five vertices. At least three faces meet at each vertex, hence

$$v \leqslant (6h + 5p)/3 \tag{A2.10}$$

The Euler characteristic of a sphere is equal to 2, so that

$$f + v - e = 2$$

Hence

$$(h + p) + (6h + 5p)/3 - (6h + 5p)/2 \geqslant 2 \tag{A2.11}$$

Therefore, $p \geqslant 12$ as required.

Scaling restrictions for the spherical tessellations

DODECAHEDRON

The area of a pentagon of radius t is $5(\sin 54° \cos 54°)t^2 \simeq 2.38t^2$. The surface area of the dodecahedron is thus $28.5t^2$. The area of the spherical time surface, for $V = \alpha r^2 + \beta$, is $4\pi/4\alpha\beta$. For these to have similar scales we require $\pi/\alpha\beta \simeq 28.5t^2$. This formula was used to calculate the parameters for Figure 3.4h.

TRUNCATED ICOSAHEDRON

The area of a hexagon of radius t is $(3\sqrt{3}/2)t^2 \simeq 2.60t^2$. The area of an adjacent pentagon of edge length t is $[(5 \tan 54°)/4]t^2 \simeq 1.72t^2$. The surface area of the truncated icosahedron, with 12 pentagons and 20 hexagons, is thus $72.6t^2$. The scaling restriction is thus

$$\pi/\alpha\beta \simeq 72.6t^2 \tag{A2.12}$$

This formula was used to calculate the parameters for Figure 3.4i.

A3   *Isochrones in the field $V(r) = \omega r^p$*

Let us consider the properties of the isochrone in the field

$$V(r) = \omega r^p \qquad (0 \leqslant p < 1) \tag{A3.1}$$

where V is the velocity, r is the distance from the city centre, and ω and p are parameters. The basic equation is derived in Corollary 1.7 of Section A1. It is

$$r_2 = \{m \cos[(1 - p)\,\theta_{12}]$$
$$\pm \sqrt{(\omega^2\,t^2(1 - p)^2 - m^2 \sin^2[(1 - p)\theta_{12}])}\}^{1/(1-p)} \tag{A3.2}$$

where $m = r_1^{1-p}$, and where r_1 is the distance of the facility from the city centre, r_2 is the distance of a point on the isochrone from the city centre, θ_{12} $(0 \leqslant \theta_{12} \leqslant \pi)$ is the angle separating the two radials, and t is the time standard. We now consider three special cases of it.

Case A   From (A3.2), for a real solution,

$$\omega^2 t^2(1 - p)^2 \geqslant m^2 \sin^2[(1 - p)\theta_{12}] \tag{A3.3}$$

Thus, at the limits A and B shows in Figure A3.1a,

$$\theta_{12} = \frac{1}{(1 - p)} \sin^{-1}\left[\frac{\omega t(1 - p)}{m}\right] = \phi \tag{A3.4}$$

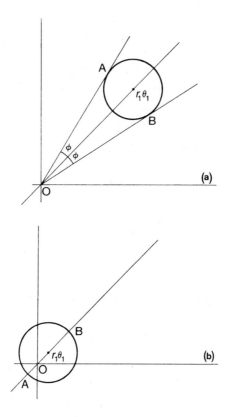

Figure A3.1 Geometric properties of an isochrone for (a) a non-central and (b) a central facility.

On the side farthest from the city centre, Equation (A3.2) is taken in the positive direction, θ_{12} ranging from $(\theta_1 - \phi)$ to $(\theta_1 + \phi)$. On the opposite side (A3.2) is taken in the negative direction over the same range for θ_{12}.

Case B   Also for a positive solution, if the negative of Equation (A3.2) is taken,

$$m \cos[(1 - p)\,\theta_{12}] \geqslant \sqrt{(\omega^2\,t^2(1 - p)^2 - m^2\sin^2[(1 - p)\,\theta_{12}])} \quad (A3.5)$$

implying that

$$r_1 \geqslant [\omega t\,(1 - p)]^{1/(1-p)} \quad (A3.6)$$

This means that the radial from an emergency centre to the city centre must cut the isochrone *en route*. If not, the isochrone cuts all four quadrants, as shown in Figure A3.1b. In this instance θ_{12} in Equation (A3.2) ranges over the interval $0 \leqslant \theta_{12} \leqslant \pi$.

Case C Suppose $r = 0$. Then the emergency centre is located at the city centre. Equation (A3.2) then simplifies to

$$r_2 = [\omega t (1 - p)]^{1/(1-p)} \tag{A3.7}$$

which is a constant for any given set of parameters. Thus the associated response area for any value of t is a circle.

4 *The demand for health services in cities*

4.1 Introduction

Thus far, some simple methods have been outlined for dividing urban areas into hospital districts. The respective properties of the methods have been analysed from the points of view of equity and efficiency, and some conclusions have been drawn regarding the relative costs of hospital provision in different parts of a city. The discussion was based on the sorts of factors that providers of health services would need to take into account when allocating health care resources. At no time during this discussion, however, were the behavioural aspects of hospital choice or utilisation analysed. For example, in the case of emergency medical provision discussed in Chapter 3, the question of choice did not arise because patients were simply taken to the nearest hospital. However, accidents and emergency cases form only a small proportion of the total demand for health care.

In this chapter, therefore, attention focuses on hospital provision from the point of view of the patient. It considers the main factors affecting demand in different areas of a city, including the relationship between hospital choice and accessibility. In particular, it examines variations in the geographical coverage of different hospital services and the possible implications in terms of service levels and resource allocation of a shifting population. A simple planning model is outlined and related to the theoretical issues discussed in earlier chapters. It shows how this model may be used for predictive purposes when there are changes in supply or demand. Finally, it is shown how the model can be reformulated to provide a more powerful tool for directing a health system towards specific objectives such as improving equity.

The discussion concentrates on the demand for acute health services. These constitute the largest sector of health care provision and are associated with the most distinctive spatial patterns of demand. If these patterns can be identified, they will assist health authorities in understanding the importance of location, both in day-to-day management and in making investment decisions. In presenting the arguments, however, a distinction is drawn between patients who are 'allocated' to a hospital in their locality (the nearest, say) and those who, after consultation with their doctors, participate in the choice of hospital. The first case indicates a strictly controlled system in which access to a hospital is determined by the patient's geography of residence; the second indicates a more informal system in which choice is determined by the interplay of factors such as hospital prestige, professional advice, accessibility and other considerations. Here, we are largely interested in the second case. Not only is it more complex, but it is more typically representative of modern health care systems.

It might be supposed that with an unlimited choice of hospitals there would be no systematic spatial patterns of provision along the lines discussed in Chapter 2. Whether in fact patterns exist depends on several factors, including the scale of the city, the number of hospitals and so forth. They in turn depend on the administrative and financial structure of the health system. If, for example, access to some hospitals is restricted on religious or ethnic grounds, or on the ability to pay, these will obviously influence the spatial patterns of demand accordingly. Plainly, these and other factors will vary from city to city so that their possible influence cannot be approached in a general way. For this reason, discussion in this chapter focuses only on the general factors that are argued to influence the geography of demand and supply.

4.2   Factors influencing the demand for health care

There are many factors influencing the demand for health care. Morbidity is the main factor and this in turn is dependent on other factors such as age, sex and socio-economic circumstances. A curious aspect of demand is that it varies considerably between different countries and regions, including those which, ostensibly, have similar characteristics in terms, say, of population and level of economic development. For example, in many developed countries life expectancy (one measure of health 'output') is broadly comparable and yet the hospital activity between these countries can vary by as much as threefold. The most plausible explanation for these large variations is not that there are particularly stark differences in morbidity. It is rather that the average population, if given the opportunity, would make much wider use of health services than they actually do. Indeed, behavioural observations of these kinds have led to one school of thought that there exists a large, latent demand for health care which is either unsatisfied in the conventional sense of using hospital-based health services or is catered for in other ways (e.g. by general practitioners, domiciliary or community services, or through self-administered treatment).

In the allocation of any form of health care resources, it is therefore important to be clear on the circumstances under which it is essentially demand that fuels supply or supply that fuels demand. Plainly, this is an empirical question which may have different answers in different places, but it is one that needs to be considered before any form of health care planning can be effective. To stimulate the debate, therefore, two hypotheses concerning the nature of demand are proposed. They are termed the fixed demand hypothesis and the elastic demand hypothesis. Both take into account the role of accessibility costs as a possible diminishing factor in the demand for health care at a particular location. The main difference is that in the first case all demand is presumed to be satisfied (if not at one particular location then at another); in the second case demand is satisfied only when there are resources available for treatment, so that the level of satisfied demand in an area becomes critically dependent on the size and location of health care facilities.

Let us consider further the elastic demand hypothesis which is plainly more difficult to interpret. One way to look at the problem is to imagine demand as being composed of essentially three separate parts. The first and most easily determined part consists of patients for whom the necessity for hospital treatment is in no doubt since without it they may die. The second part, by contrast, consists of patients for whom hospital treatment is still essential but the nature of their clinical condition means that treatment can be deferred – perhaps indefinitely. Finally, the third part consists of prospective patients who may or may not be ill and who could, if they chose to, obtain treatment or consultation somewhere other than in a hospital. According to the elastic demand hypothesis it is the second and third components of demand that are the most elastic in the sense that they are the first to be affected when the supply of health services in a region or locality is increased or reduced. Implicit in this behaviour, of course, would be the assumption that health authorities operate a scale of priorities that, depending on the availability of resources, prevent the services from being either overstretched or under-utilised.

4.3   Spatial patterns of demand

In this section we turn to the identification of the basic spatial patterns of demand for acute health services in cities. The more difficult question of how these patterns are affected when certain variables are assumed to alter, depends, of course, on which of the above hypotheses – fixed demand or elastic demand – is considered. This problem is discussed later from Section 4.6 onwards, when attention focuses on the planning model. To begin this section, we turn to Figure 4.1, which shows two hypothetical cross-sections of demand through two cities in which (a) there is a negative expotential decline in population density from the centre to the periphery of the city. From this diagram a lot of general information is conveyed concerning the spatial patterns of demand for different types of services and how these are affected by the concentration of population. In case (a) the locations of the hospitals, denoted L_1, L_2, . . ., etc., are spaced at equal intervals, according to the districting principles described in Chapter 2. Likewise, in case (b) the intervals reflect similar considerations; however, because the population density is non-uniform and concentrated at the centre of the city the intervals vary becoming wider towards the edge of the city.

Curves representing the density of demand for services at each location are of two types. The taller, more steeply declining curves are proportioned to the demand for a locally based service where the volume of demand is high. The

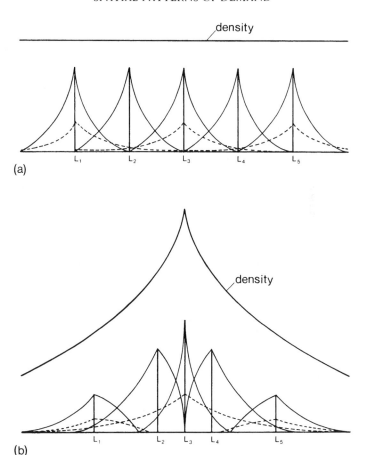

Figure 4.1 Cross-sections of demand through two cities: (a) with a uniform distribution of population and (b) with a negative exponential distribution of population. The curved profiles show the hypothetical variation in the demand for different hospital services in the locality of a hospital. ——————, local service; — — — —, specialised service.

flatter, gently declining curves are proportioned to the demand for a more specialised service. In the diagram, it is seen that this service is available at only three of the five hospitals. It is noteworthy that in both case (a) and (b), none of the curves are drawn so as to terminate at the boundaries separating the hospital districts as defined in Chapter 2. This is indicative of a very important behavioural effect which we have not so far considered, namely the existence of 'cross-boundary' patient flows between adjacent districts. The magnitude and extent of these flows depend not only on the specialised nature of the services concerned, but also on the spatial intervals between the hospitals. Thus, in case (b), hospitals near the city centre have closer proximity and the degree of overlap is depicted to be greater. We may usefully contrast this portrayal of spatial behaviour with the assumption in central place theory (Sec. 2.5) that there exists a finite outer range of travel for each category of service which is related to the service threshold. It may be noted in passing that the mathematical requirement for a service to be provided in the new case is that the integral under the demand curve for a particular service is equal to or larger than the threshold for that service.

A further significant property of these patterns applies only to case (b): the shape of the demand curve in the case of a non-centrally located hospital on the side nearest the city centre is different from the shape of the curve on the side facing away from the city centre. Specifically, the profiles of demand facing the centre curve outwards and those on the far side curve inwards. It is particularly noteworthy that the latter are also characterised by a long 'tail' stretching into the suburbs, indicating the typically longer average distances that patients originating in these areas tend to travel for treatment. The only location, L_3, where these differences do not apply is at the city centre, where the demand attracted by a hospital is evenly balanced on both sides. This will always be the case providing the population is symmetrically distributed around the city centre.

4.4   Details from an outpatient survey

At this stage, the patterns of demand described above are still speculative, and they need to be compared with suitable data before proceeding further with the discussion. However, this is not easy due to the lack of appropriate information at the correct geographical scale. This is because health authorities assemble data on the basis of administrative areas, which are spatially too coarse for a study of micro-spatial variations in demand of the type currently under consideration. Some pointers are possible, however, from a small survey of hospital outpatients in the London area carried out and briefly analysed below as part of a broad validation of the proposed patterns (Mayhew 1979). In analysing outpatients (and not inpatients) it was implicitly assumed, however, that the spatial patterns so derived would be representative of the spatial patterns of other hospital users. In fact, this is only an approximation since a particular outpatient clinic may not have inpatient facilities in the same hospital.

In all, data were collected on about 2000 patients at 14 London hospitals. The information obtained included details of where patients lived and started their journey to hospital and the clinic they were attending. The data were separated into two clinical groupings, the first consisting of surgical and general medical outpatients and the second, a less homogenous and more specialised group, consisting of the remaining clinics. Additional details are given in Mayhew (1979).

There are two main issues arising from the survey of relevance to the present

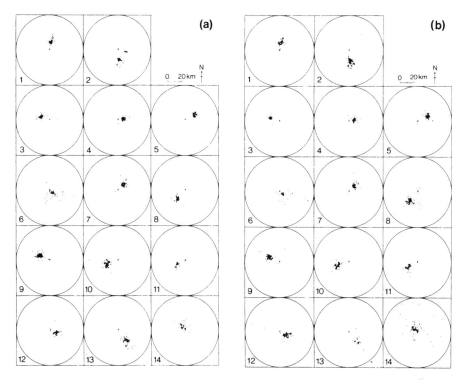

Figure 4.2 Maps showing clusters of demand for (a) category 1 and (b) category 2 outpatient clinics at 14 London hospitals.

discussion: (a) whether the data indicate differences in the patterns of demand on the sides of each hospital nearest and farthest from the city centre and (b) whether the choice of hospital is related to the location of the hospital within the urban area, to the type of clinic and to the relative proximity of other hospitals. To assist in the evaluation of (a) and (b) Figure 4.2 shows maps of the clusters of demand around each hospital spread at different points of the compass for the two clinical groupings defined above. The hospitals concerned are located at between roughly 5 and 20 km from the city centre.

Patterns of demand in the vicinity of a hospital
Figure 4.3 shows two representative histograms indicating the percentage of patients originating at different distances *either* side of two of the hospitals (numbers 4 and 7 in Fig. 4.2) in which the data for clinical categories one and two have been combined. In both examples it is seen that the peak level of demand occurs, as predicted, on the side nearest the city centre, very close to the hospitals. On the side farthest from the city centre the histograms are flatter and have the characteristically long 'tail', also predicted in Section 4.3,

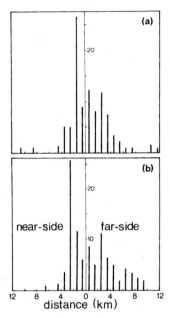

Figure 4.3 Variations in the percentage of patients originating on the near side and far side of two hospitals in London: (a) hospital 4 and (b) hospital 7 in Figure 4.2.

indicating that patients originating in the suburbs typically travel farther. If we now define the focal centre of demand $(\bar{x}, \bar{y})$ as the average of the x, y coordinates of each patient's home or journey origin, then for nearly every hospital in the survey the results show this focal centre is located on the far side of the facility. This finding is hence another indication of the 'pull' effect that hospitals have on demand originating at the edge of the city. Among the hospitals evaluated, the one exception to this behaviour was a hospital whose cluster of points was truncated by the River Thames, which flows through the city. This case would thus represent a typical example of a spatial 'inefficiency' of the type considered in more detail in Chapter 3, because access to the hospital in question is reduced by a physical barrier. One final point to emerge from the analysis of these histograms is the way the average distance travelled changes relative to the distance of each hospital from the city centre. This average is stable at about 2 km for hospitals located up to 10 km from the centre. After this it increases rather rapidly to about 6 km, as the clusters start to exhibit more radial stretching. It is of interest to note the correspondence between this observation and the theoretical districting pattern in Figure 2.4, which shows a similar stretching effect in suburban areas of the city.

The extent of cross-boundary flows
According to Section 4.3, two factors influencing the choice of hospital and hence the potential extent of cross-boundary flows are the distance of the

hospital from the city centre and the category of clinic. In particular, the more specialised the clinic, the more the clusters of demand points should encroach on the locations of neighbouring facilities since the chances are that they will be offering only the less specialised clinics on an outpatient basis. A simple method of characterising the extent of encroachment is to calculate the percentage of patients in each clinical category that use their nearest hospital. If this percentage is high, the district is relatively self-contained; if it is low then there is a considerable overlap with the districts of neighbouring hospitals. An analysis of the first clinical category, for example, shows that, for hospitals nearest the city centre, the percentage is about 10 per cent, rising to 70 per cent at 20 km from the city centre. The measure of correlation between this percentage and the distance of the hospital from the city centre is 0.62, suggesting that the amount of self-containment is strongly dependent on the location of the hospital with respect to the city centre. Selecting individual clinics from within clinical category two, and comparing the spread of distances for category one patients, indicates further systematic effects. For example, patients attending a clinic for kidney and liver conditions at one of the hospitals travel substantially further than patients attending a gynaecology clinic at the same hospital, who in turn travel further than general medical patients. Such rankings based on distance travelled, however, must be considered only as indicative because of the strictly limited scope of the survey. Nevertheless, they should be regarded as demonstrations of the patterns of spatial choice consequent upon the availability of different services in different locations.

4.5   The impact on demand of a shifting population

Whereas these illustrative pointers were based on snapshots of particular locational patterns, hospital authorities are likely to be more interested in what happens to these patterns when there is a shift in population and hence demand. The objective of this section is to contrast the possible consequences resulting from such shifts when one or other of the two demand hypotheses discussed in Section 4.2 is adopted. In Section 4.7 evidence is presented for selecting one of these two hypotheses for use in a planning model. The arguments for the moment, therefore, continue to be speculative.

Fixed demand
The fixed demand hypothesis assumes that the population makes use of health services up to its needs and that it is directed to facilities partly on the basis of their accessibility and partly according to their attractiveness as medical centres. Figure 4.4 illustrates changes in the hypothetical levels of demand for one particular case which is representative of the problems in which we are interested. This concerns the impact on the supply of services following a

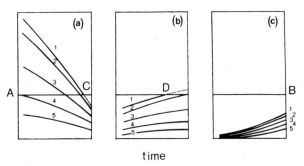

time

Figure 4.4   Changes in total demand over time for different services at (a) 0 km, (b) 5 km and (c) 10 km from the city centre on the basis of the fixed demand hypothesis; the horizontal line AB indicates the threshold level for service 1.

deconcentration of population (and hence demand) at three locations in a·city which has a centre to periphery decline in population density. The three locations are at the city centre, in an inner suburb, and in an outer suburb. On the vertical axis of the three graphs is demand where the measurement units are cases per annum; on the horizontal axis is time, where the period under consideration is of the order of 50 years. Consider initially demand at the city centre (Fig. 4.4a) where the effects of population decline on service provision are expected to be most pronounced. The service showing the steepest fall in demand is service 1, a local service with a high case-load. The service showing the least fall in demand is service 5, a specialised service catering for a much wider area of the city and less affected, therefore, by a loss in local population. In the inner and outer suburbs (Fig. 4.4b, c) the inflow of population has the opposite effect, with demand for service 1 increasing proportionately more than the demand for service 5. Because the inflow of population occurs slightly later, demand in the outer suburb lags slightly behind demand in the inner suburb.

Plainly, one general implication of these results is that hospitals located near the centre would reduce in size, while those in the suburbs would increase in size. An important qualification, however, is that the rate of growth or decline in each service would differ. This is an important problem for health authorities, which have to decide where and when to adjust the level of service provision in each location. One approach for taking such decisions is to use the threshold concept introduced in Chapter 2. A threshold, it will be recalled, represents the critical level of potential demand, before a service is eligible for provision in a particular location. If the threshold is presumed, for simplicity, to be constant over time and is measured in cases per annum in diagrammatic terms, it can be simply represented as a horizontal line, such as AB in Figure 4.4. As an illustration, suppose the threshold indicated by AB refers to service 1, then this service is withdrawn when demand cuts AB at point C from above (Fig. 4.4a), and is introduced when it cuts AB at point D from below (Fig. 4.4b). Note that this account is highly simplified because the level of demand attracted to each facility would partly depend on whether the threshold is satisfied at other facilities. However, similar principles would apply in the more realistic case.

The elastic demand case

In the second category of demand behaviour, the demand for a hospital service rises to meet the level of resources allocated. For example, suppose the capacity of a hospital near the city centre remains constant but the local population declines as a result of suburbanisation. Then, according to the hypothesis, this will result in an increase in hospitalisation rates locally. Conversely, if the local population increases, then local hospitalisation rates are reduced. Because different services are effective over different areas the extent of the increase or decrease in the hospitalisation rates for each service will also vary. The amount of variation depends critically on the spatial range of the service in question. Thus, if the change in population is very localised, then the effect, of course, will be mainly on the local services. Note that the use of the threshold in this instance is slightly different because the question of whether demand is above or below the threshold does not arise (since all resources are used). In this instance its application would be to co-ordinate the levels of supply in each area so that the resources allocated to different services reflect the relative needs of the area and the economics of providing the services in question.

4.6 Some options for health authorities

A fundamental question arising from the above discussion concerns the strategies that are open to health authorities to enable them to cope with population change and the difficult problem of correcting resultant imbalances in supply and demand. Surprisingly perhaps, the options are similar regardless which of the demand hypotheses – fixed or elastic – is chosen. The main difference lies in the speed with which hospital authorities are likely to respond to each hypothesised mode of demand behaviour. In particular, the likelihood is that the response, in terms of adjusting service levels, will be faster in the fixed case because the impact of falling demand will be immediately felt at the points of supply. In the variable case it may take time before health authorities detect any imbalance in service provision since, if the hypothesis is correct, there will be no immediate noticeable effect on demand (although there may be an effect on the types of cases being admitted to hospital). In this instance, it would normally take significant disparities in hospital usage in different parts of the city to precipitate any remedial action.

The most effective, but also the most expensive solution for redressing any imbalances is simply to re-locate the hospital and its constituent services to areas where there are the biggest shortfalls in provision. A major difficulty, however, is choosing the location which optimally integrates the new hospital with existing services in the area and ensuring alternative provision for the residual demand in the old location. Instead of relocation, a second and obviously cheaper approach is to replace marginal or declining services with others that attract either more patients or patients from a larger area, or which fill particular or emergent gaps in the services already available. Flexibility in staffing and equipment is needed, however, for success with this strategy and, for these reasons, it tends to favour larger rather than smaller hospitals. A third approach is to change completely the functions of the hospital from, say, providing acute care to providing care for the elderly or chronically sick. Whether this option is viable, however, depends on several factors, including the suitability of the premises and their location in relation to the constituent demand for the new functions. If it is viable, then this option is an important way of extending the useful life of hospital buildings and equipment. The fourth and final option, of course, is hospital closure. In summary, therefore, the problems of rearranging services to cater for a large and shifting population are both numerous and complex. In the next section, we consider a simple planning model which is finding increasing application in the evaluation and planning of hospital services precisely to solve these problems.

4.7   An outline of a spatial planning model for acute health services

Thus far, the basic patterns of demand for acute health services in cities have been analysed, and some implications of the impact of a shifting population on health services in different locations have been considered. Clearly, it would be particularly useful if health authorities were able to predict in advance the broad impact of some of the changes described. This would not only facilitate a better forward planning of health services but it would also help to pinpoint possible gaps in provision. More importantly still, it would also facilitate the implementation of policies designed to improve the equity or efficiency of a health care system. The remaining sections of this chapter therefore outline the formulation and application of a simple mathematical planning model that addresses some of these issues. Because the technical details of the model are, in places, lengthy to describe, the outline here makes certain simplifications. Firstly, it does not consider acute services separately or the various possible interactions between them but rather it considers services in aggregate. An appropriate model for the more complicated case of multiple services is analysed in Mayhew and Leonardi (1984). Secondly, it does not examine detailed statistical issues relating to questions of data availability or to the validation of the model. These are discussed, for example, in Mayhew and Taket (1980, 1981), Hall (1982), Tadei et al. (1983), Rising and Mayhew (1983), and Mayhew et al. (1985).

The first problem in formulating the model is to decide which of the two

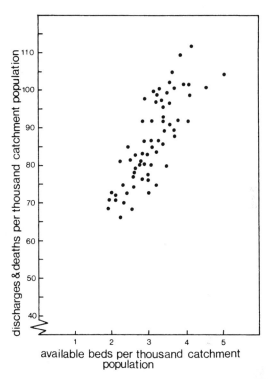

Figure 4.5 A scattergram showing the relationship between cases treated and available beds per thousand catchment population in south-east England in 1977 for acute hospital services; correlation coefficient = +0.85.
Source: LHPC 1979.

demand hypotheses is most appropriate in a particular situation. This is not a trivial question and it requires a careful investigation of the patterns of use of hospital facilities in a particular city or region. Consider first Figure 4.5, a scattergram indicating an important empirical relationship based on the use and supply of acute hospital services in London and the surrounding area. On the vertical axis is the number of deaths and discharges per thousand catchment population (one measure of demand); on the horizontal axis is the available number of beds per thousand catchment population (one measure of supply). In the constituent areas of this region, there are known to be considerable variations in health care needs due to differences in the population structure and other factors. The scatter of data points, by contrast, suggests that the demand variable seems to be closely related to the supply variable (correlation coefficient, 0.85). The inference therefore, is that *within* the range of data values shown the elastic demand hypothesis appears to be more appropriate than the fixed demand hypothesis. What happens outside this range is, of course, an open question, and must therefore be regarded as uncharted territory as far as the model is concerned. Although uncharted areas of demand and supply are of intrinsic interest in the formulation of government policy towards health care, practical circumstances dictate that health authorities will be more interested in change that occurs within the observable range. This, then, is the domain principally addressed by the model and the remainder of the discussion. If we had wanted to use the demand–led model instead, we would first need to establish the demand for hospital treatment independently of the resources available. This means that we would also have to account for the significant variations in demand shown in scattergrams like that in Figure 4.5, and this is by no means an easy task.

The gravity hypothesis underlying the proposed model
The behavioural hypothesis on which the model depends is known as the gravity hypothesis. According to this hypothesis the flow of patients between a place of residence i (a town, local authority, etc.) and a place of treatment j (a hospital, health district, etc.) is proportional to the expected demand or relative need for health care in i and the resources available in j, but is inversely proportional to the accessibility costs (distance, travel time, etc.) of getting from i to j. Models based essentially on the gravity hypothesis have been used for many years to analyse the spatial aspects of a variety of different activities, including shopping and recreation patterns and journeys to work (e.g. see Wilson 1974). Which type of model is used depends, as in the case of health care provision, on the underlying behaviour of the system. Two particular models in wide usage fall neatly into the current classification of demand behaviour considered here. These are the demand–led model and the supply–led model. For example, the model most frequently used to plan and analyse retail activities is a demand–led model, and it is summarised in the appendix to this chapter. As far as this discussion is concerned, however, we have opted to use the supply–led version. In practice, the main difference between the models is that a constraint variable is attached to the supply–led model which prevents the number of cases treated at hospitals in an area from exceeding their case–load capacities. In the alternative, demand–led version a similar constraint ensures that the total number of cases treated cannot exceed the total demand in the areas of residence. The reason why this version finds more favour in the planning of retail services is because retail systems are essentially uncapacitated and customer waiting times are relatively short. Hospitals, by contrast, have essentially finite capacities determined by the number of beds available for treatment. It is this consideration as well as statistical evidence which recommends the use of the supply–led model in health care systems. In writing the supply–led model mathematically, a standard notation is adopted based on the paper by Mayhew and Taket (1980), in which some of the symbols duplicate variables already used in Chapters 2 and 3. Since the model is considered only in this chapter, no confusion should result.

Statement of the model
Mathematically, the model can be stated as follows:

$$T_{ij} = B_j \, W_i \, D_j f(\beta, \, c_{ij}) \tag{4.1}$$

where T_{ij} is the predicted flow of patients between areas i and j; D_j is the case-load or 'treating' capacity of hospital facilities in j for a specified group of acute services; W_i is the relative health care needs or 'patient generating potential' of area i; c_{ij} is the accessibility costs of getting from i to j; $f(\beta, \, c_{ij})$ is a monotonic declining function of accessibility costs c_{ij} and β, a model parameter (later the function will be abbreviated to f_{ij}); B_j is the constraint variable which ensures that the number of cases treated does not exceed the case-load capacity of hospital facilities in j. More precisely, this variable ensures that

$$\sum_i T_{ij} = D_j \tag{4.2}$$

It is easily shown that this always satisfied when

$$B_j = \left[\sum_i W_i f(\beta, \, c_{ij}) \right]^{-1} \tag{4.3}$$

The specification of variables
Before the above model can be used for planning purposes, it is necessary to specify in more detail the nature of the variables and the data they use. The variables in question are the case-load capacities, the patient generating potential and the accessibility costs.

Case-loads   In the model, supply is specified in terms of the case-load capacities of the hospitals. Case-loads are related to other measures of supply, such as hospital beds (see also Ch. 2) by the following simple expression:

$$\text{available bed-days per year} = \text{beds} \times 365 \tag{4.4}$$

or

$$\text{available bed-days per year} = \text{cases treated} \times (l + t) \tag{4.5}$$

where l is the average length of hospital stay and t is the average turnover interval between successive cases. It thus follows directly from these definitions that

$$\text{number of beds} = \frac{\text{cases}}{365} (l + t) \tag{4.6}$$

One implication of this relationship is that we can use the model to predict the impact on demand of adding or subtracting different numbers of hospital beds in different locations. In using the model in this way, however, it is implicitly assumed that the variables l and t are relatively stable as compared with the case-load. The first variable, the average length of stay, is related to case severity and other factors, and the second variable to admission and discharge patterns. In practice, it may be reasonable to make this assumption, but if these variables tend to have a significant impact on the results then their effects must be taken explicitly into account. Methods exist for analysing changes in lengths of hospital stay (e.g. see LHPC 1974), but we do not discuss these here.

Patient-generating potential   In the supply-led version of the planning model the use of the term demand is perhaps anomalous because, if the hypothesis is correct, demand is an elastic quantity which depends on the supply of resources available. The term patient-generating potential is possibly a better description in that it is indicative only of the relative (not absolute) demand in each area.

The simplest measure of relative demand is population and it will be recalled that this formed the basis for one of the districting criteria in Chapter 2, namely the P-criterion. One possible refinement is to weight the population in each area according to the relative propensity of different age and sex groups to use hospital services. Another is to introduce socio-economic considerations which are believed to affect the relative need for hospital services. An example of the first refinement is written mathematically as follows:

$$W_i = \sum_l \sum_k P_{il} U_{lk} \tag{4.7}$$

where P_{il} is the population in area i and age–sex category l and U_{lk} is the national (or regional) utilisation rate in service k by age and sex. It is noteworthy that this expression also has another interpretation. It is also the expected level of demand in area i based on national (regional) patterns of hospital utilisation. Plainly, this is not necessarily the same as the demand which is actually satisfied; in the model that is determined on the basis of accessibility to supply. An example of the second type of refinement, based on socio-economic considerations, is to weight W_i by factors such as relative income or mortality. How far to proceed with these modifications, however, is essentially an empirical question. The general principle is to develop a measure that is closely representative of the sub-group of the population that might avail itself of a particular service if it were made available. For example, if we were concerned only with certain types of day surgery then we would select only those surgical categories of treatment which lead to a quick recovery. Another refinement would be to take into account the availability of non-hospital based treatment facilities in an area because if very large, it may have a reducing effect on the patient generating potential. However, further research is needed on this issue.

Accessibility costs The third variable is the accessibility costs, c_{ij}. These are specified in terms of distance, travel time or some weighted combination of both. The objective is to determine them so that they reflect the way 'distance' is actually perceived by patients. In the model, the accessibility costs are important because they modify the function, f_{ij}, which is sometimes called the deterrence function. According to the hypothesis, patients are more 'deterred' from using hospital facilities if accessibility costs are high than if they are low. The other component of f_{ij} is the model parameter β which is described in more detail below. A typical mathematical form of the deterrence function, for example, is the negative exponential function, which is written as

$$f_{ij} = e^{-\beta c_{ij}} \tag{4.8}$$

Another functional form for f_{ij} in common use is the power function. In addition, it is possible to infer the functional form directly from data on patient flows (see below and the appendix to this chapter).

Correspondence between the model and urban demand profiles If we return briefly to our theoretical profile of demand shown in Figure 4.1b, it is simple now to show the connections with the planning model. Consider Figure 4.6, showing a portion of the original profiles which has only one hospital located in a non-central position indicated by j. Imagine now the places of residence i shrinking in size, but multiplying in number, so that they maintain their coverage of the urban area without any gaps. Then imagine a cross section through the demand generated in an area in the immediate vicinity of the hospital shown in Figure 4.6. In particular suppose that the relative demand (i.e. W_i in the model) varies at different distances from the city centre such that a portion of it can be presented by the smooth curve AB. The curves CD and

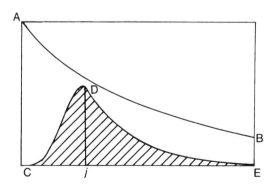

Figure 4.6 Profile of demand in the vicinity of a non-centrally located hospital as indicated by the shaded area, where j denotes the location of the hospital and AB the local variation in potential demand.

DE then represent the actual level of usage made of the hospital at j by residents in each i (now reduced to an arbitrarily small size). If the model hypothesis is correct the integral of the shaded area under these curves will equal the case-load capacity of the hospital.

The model parameter β controls the exponential rate of descent of the shaded curves. In the current discussion we have been considering all acute services as though they were one and the same. If individual services are considered, then different values of β would be appropriate for different services (Rising & Mayhew 1983). Specifically a small value of β would result in a much flatter curve and therefore a more specialised service than a case with a large value of β. This provides, therefore, another link with the theoretical profiles of demand in Figure 4.1, because the heights and slopes of the profiles were seen to distinguish between different services. Another refinement is to define a different value of β for each hospital. The theory is that hospitals of the same size offering similar services may differ in their attractiveness and this is one way of reflecting this in the model. One final link that needs underlining is the relationship between the shapes of the demand curves on the near side and far side of j. For the former to be more compact and to exhibit a steeper descent, as in the theoretical case, it is sufficient in the planning model for the cost of travel per unit of distance on the near side to be more expensive. It will be recalled that this is consistent with the discussion in Chapter 3 which examined the spatial impact of variations in travel time in cities.

Calibration and prediction   The above explanation shows, in intuitive terms, one way of proceeding from the hypothetical abstractions discussed in the introduction to the development of a practical planning tool based on a supply-led gravity model. Regardless of which type of model is used, however, it is important to distinguish between two different model procedures: calibration and prediction. Calibration, for example, is the statistical procedure for selecting a value of the model parameter β such that the actual flows of patients during the base-year are adequately reproduced by the model. A combination of maximum likelihood and regression techniques may be used for these purposes (see Mayhew & Taket 1980). Prediction, by contrast, is the process of modelling change by systematically altering the variables of supply and demand. The behavioral assumption is that the model parameter remains constant over the period in which the predictions are made. In some cases, depending on the availability of data, the process of prediction is greatly simplified by the use of an inferred deterrence function which allows the user to omit the calibration stage. Details are given in the appendix to this chapter.

In using the model for predictive purposes, future estimates of the supply and demand variables are made on the basis of demographic forecasts and the likely growth in case-loads (Mayhew 1980). These in turn are related to trends in hospital utilisation in different services and lengths of hospital stay. For example, in some services lengths of stay are declining, indicating an increase in throughput and hence potential case-load capacity. Having established these estimates the model can then be used to examine the consequences of varying the case-loads in different areas. Included in the outputs of the model are the predicted flows between areas of residence and treatment, hospitalisation rates by area and the catchment populations of the hospitals. Catchment populations are of particular interest because they suggest another connection between theoretical districting concepts of Chapter 2 and the planning model. The difference is that, in the model, catchment areas overlap and are modified automatically when the relative locations of hospitals are adjusted. Overlapping catchment areas, however, are more difficult to conceptualise than non-overlapping districts which is why hospital authorities, when planning services, usually prefer to think in terms of catchment populations. If we define E_{ij} as the ratio of the proportion of cases in i receiving treatment in j to all cases in i receiving treatment in all j, we can write

$$E_{ij} = T_{ij} / \sum_j T_{ij} \qquad (4.9)$$

Now, the catchment population of j is defined as the sum of the populations in all i who are dependent on hospital facilities in j. Since Equation (4.9) is the proportion of all cases in i using facilities in j, we can also write

$$C_j = \sum_i E_{ij} P_i \qquad (4.10)$$

where C_j is the required catchment population of hospitals located in j.

One further issue we should mention concerning the interpretation of the results is the question of the extent to which demand reduces in remote areas following the closure of their only hospital facilities. Because of the very high accessibility costs to other areas with alternative facilities, the model could predict very large reductions. This is actually a useful feature of the model because the user would be quickly alerted to the possibly harmful social effects of such a closure policy. In the event that such results are considered unreasonable or do not accord with experience obtained closing other remote facilities, it is useful to consider again our original characterisation of demand, which was that it consisted both of an elastic component and an inelastic component. The latter, it will be recalled, consists of the urgent cases who, it is argued, will always receive the appropriate treatment. Now, total demand, as predicted by the model, increases steadily with W_i, the patient generating factor, levelling off towards a point determined by the availability of treatment facilities. The inelastic or urgent component of demand, by contrast, tends to rise in direct proportion to W_i, as seems intuitive, because a given population is associated with a relatively constant number of urgent cases. If supply were constrained, it follows that, for a sufficiently large value of W_i, the line of minimum demand will eventually cut the curve of total demand, causing a discontinuity. At this point all admissions to hospital from the remote area in question will comprise only urgent cases. If we were to include this behaviour in the model, then adjustments would need to be made so that the appropriate minimum level of demand is always satisfied. Fortunately, in densely populated areas, the level of supply is such that this refinement is not ordinarily required, although it might be a consideration in poorer societies where the availability of health care resources is more constrained. More details on this particular refinement are given in Mayhew *et al.* (1985).

4.8   Towards the more effective planning of health systems in cities

The above examples are representative of the useful information available from the planning model and how they inter-relate with the elastic demand hypothesis and districting concepts. One problem faced by health authorities using the model, however, is the extremely large number of alternatives the model can simulate. In considering these alternatives health authorities will usually be more interested in those which are consistent with wider, strategic policy objectives such as achieving a more equitable or efficient balance of resources within the region. Chapter 2, for example, considered such issues in the context of different theoretical spatial arrangements of hospitals. While these theoretical patterns provided useful insights into the economies of hospital provision in cities, they could not be used directly for forward prediction. The interesting question, therefore, is whether the planning model can be reformulated in a way that provides a practical method of narrowing down the set of alternatives to much smaller set of alternatives which are consistent with particular policy goals.

Equal access for equal risk
To illustrate the concepts involved we choose one particular objective which has achieved a broad acceptance in health services planning. This is to provide equal access to services for those equally at risk of hospitalisation. Plainly a desire to improve equity is the motivation underlying this objective, but for it to provide the basis for a realistic set of planning options several factors first have to be taken into account. These factors can be grouped under two broad headings: systems constraints and economic constraints. The first reflects the overall availability of resources for allocation and the practicality of transferring those resources from one area to another within a given planning horizon. For example, there may not be sufficient land or finance capital available to make the set of changes needed in the time available, in which case the next best alternative must be selected. There may also be a need to maintain in an area a minimum provision for reasons of remoteness from other areas or for training and teaching functions whose requirements may conflict with the service needs of the population. By varying the constraints and re-running the

model, each option can, in principle, be tested in the knowledge that the underlying objective of equalising access is being taken into account.

The second set of constraints are economic and they basically represent the threshold for each service. In the present discussion, we are considering an aggregate model comprising all acute services, so that the threshold in this application is, strictly speaking, inappropriate. Normally, however, it would operate in the following manner. Resources would be allocated by the model on the basis of the given strategic objective (in this case equity). The allocations would then be examined to see whether they satisfied the threshold. If some do not, the services would be withdrawn sequentially from each location concerned and the model would be re-run over the reduced set of locations, assuming the same total quantity of resources. This process continues until the allocations satisfy the threshold criterion at the reduced set of locations. At least one location will always receive an allocation providing the total resources available for allocation exceed the threshold. Clearly, if both the systems constraints and the economic constraints are operating even more flexibility is permitted. However, their combined use can lead, in special circumstances, to contradictory locational decisions. For a full treatment of these issues, see Mayhew and Bowen (1984).

Details of the equity procedure
The first step of the equity procedure is to take the original model,

$$T_{ij} = B_j W_i D_j f_{ij} \qquad (4.1)$$

Then, by summing the variables over j, we obtain

$$\sum_j T_{ij} = W_i \sum_j B_j D_i f_{ij} \qquad (4.11)$$

The left-hand side of Equation (4.11) is now the predicted number of patients living in area i treated in all hospitals located within the region of interest. Dividing both sides of this equation by W_i, the patient-generating factor, we have

$$(\sum_j T_{ij})/W_i = \sum_j B_j D_j f_{ij}, \qquad (4.12)$$

which is also the ratio of patients receiving treatment in i to the *expected* number of patients needing treatment. For a given allocation of resources, the equity principle requires that this ratio is constant in each area of residence, because each person at risk should have an equal expectation of receiving treatment; that is,

$$(\sum_j T_{ij})/W_i = \alpha = \sum_j B_j D_j f_{ij} \qquad (4.13)$$

If the total resources available exactly match expected needs, then α will equal unity. Otherwise, depending on how W_i and Q are specified (see below), it will differ. Consider now the task facing health authorities, which is to share out resources so that the equity objective obtains throughout the city. Mathematically speaking, the problem may be formulated as a quadratic programming problem as follows:

$$\min_{D_j} \sum_i (\sum_j B_j D_j f_{ij} - \alpha)^2 = Z \qquad (4.14)$$

This function, Z, requires us to select a vector of D_j's, the case-load capacities of the hospitals (i.e. the resources), so that the difference between the ratio of those receiving treatment and those needing treatment is as close as possible to α in every area of residence. Next consider the systems constraint. This ensures that the allocated case-loads add to the total case-load or 'budget' for the region; that is,

$$\sum_j D_j = Q \qquad (4.15)$$

The lower and upper limits on case-load allocations, which constitute the third element of the problem, are then written as follows:

$$D_{j(min)} \leq D_j \leq D_{j(max)} \qquad (4.16)$$

where $D_{j(min)}$ and $D_{j(max)}$ are the upper and lower limits respectively. Briefly, between these constraints and subject to certain technical qualifications the model selects a set of allocations which is the most equitable possible. Having selected the appropriate case-loads, the bed consequences of the predictions can then be calculated by applying the formula in Equation (4.6). (For further technical details of the above procedure see also the appendix to this chapter.)

An illustrative example
A simple illustration of applying the model as a tool for investigating the equity issue will suffice to underline the power of the approach. The example concerns the area of London currently within the administrative border of the Greater London Council (GLC), containing about 7 million residents. At the time of applying the model, this area was divided into 33 administrative boroughs (i.e. local authorities), defined as the places of treatment. Note that a health district here represents an administrative unit and should not be confused with the districting concept described in earlier chapters. Note in addition that we could have used data pertaining to each hospital rather than working at the slightly more aggregate level of the health district which usually contains several hospitals. The reason for this was to reduce the size of the model to manageable proportions for the problem then under consideration. In 1977 the hospitals located in health districts inside the GLC

treated over 900 000 patients living within the same borders, in 23 acute hospital services (for a list see Mayhew & Taket, 1980). The present illustration is designed to show the sorts of changes that would have needed to have taken place in the allocation of resources in the constituent health districts of the same area in order to promote a more equitable pattern of services for the residents, given the 1977 case-load capacity (i.e. 900 000) and the 1977 distribution of relative demand.

To retain clarity of interpretation, the lower and upper limits on allocations ($D_{j(min)}$ and $D_{j(max)}$) have been fixed at arbitary levels. Specifically, the former are set at 75 per cent of 1977 values so that no area is permitted to lose more than 25 per cent of its then existing case-load. The upper limit on allocations in each j is set deliberately to an arbitrarily large value so that the areas with the greatest shortfall in case-load capacity can be identified. In practical applications far more consideration, of course, would need to be given to these limits, fixing them individually. For example, it is taken for granted that their determination would involve extensive consultation with health authorities, patients representatives, medical and other opinion. The idea would then be to use the model in a 'what if' manner, varying the constraints iteratively.

There are three main ways of analysing the results of the model which we will now briefly consider. They are to examine the changes in the resources allocated to each place of treatment, to evaluate the changes in the numbers of cases generated in each area of residence which are predicted to arise as a result of the reallocations, and, finally, to measure the consequent change in equity. One simple and effective method of evaluating the latter, for example, is to create a scattergram showing, on the vertical axis, the predicted number of cases by area of residence against, on the horizontal axis, the relative needs of each area of residence. According to the theory, if the predicted cases in each area correspond exactly with relative needs of each area all points in the scattergram would lie on a straight line. The angle of the straight line would be $45°$ to the origin of the graph, indicating that satisfied needs are in exact proportion to relative needs. If the constraints are operating, then pure equity is unobtainable in which case the scatter of points is disturbed. The degree of scatter obtained in these cases is hence one measure of the consequent loss in equity.

In order to facilitate the interpretation of the results, Figure 4.7 shows numbered maps of the places of residence and treatment in the GLC area. The maps differ because, in this region, local authority boundaries are not the same as health district boundaries; in other cities they may be the same in which case the distinction will be irrelevant. Figure 4.8 shows two bar-charts of the changes in resource levels and cases generated by area as determined by the model. These refer to the first two methods of illustrating the outputs. Figure 4.9 shows two scattergrams of the type described above showing the cases generated versus relative needs before and after the reallocation of resources.

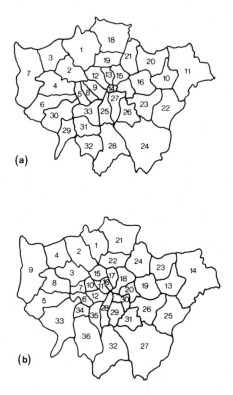

Figure 4.7 Two maps of the Greater London Council showing (a) local authority areas defined in the model as places of residence and (b) district health authorities defined as places of treatment.

Consider initially the changes shown in the bar-charts and their correspondence with the maps. A close examination shows that, to achieve a more equitable balance, resources need to be transferred from the centre of the region to the boundaries of the GLC area. Indeed, several of the inner area health districts lose the full permitted 25 per cent of their current allocations. Of the districts gaining from the reallocation procedure, districts 14 and 23 are perhaps the most notable because their allocations are doubled, suggesting that new or substantially enlarged facilities might be required. Thus one principal conclusion from this illustrative application of the model is that the centre of the region is relatively over-provided with resources and the periphery is relatively under-provided. Later, in Chapter 5, we return to these questions of disparities in provision, but from another standpoint.

Turning now to the issue of equity, Figure 4.9 shows the resultant change in equity that would occur if the resources were reallocated according to the model's predictions. As is seen there is a dramatic change in the scatter of points. In the first case, the correlation coefficient between cases generated and

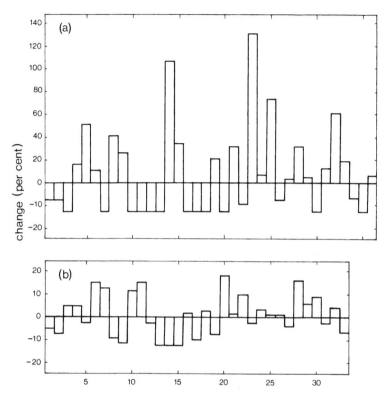

Figure 4.8 Bar-charts showing (a) the predicted change in resource levels by place of treatment and (b) the predicted change in the numbers of patients treated by place of residence.

relative need is 0.83 whereas in the second case the correlation is 0.99 – a significant improvement; but noticeably it is still slightly less than 1.00, the maximum theoretically obtainable. This is due to the effects of the lower limits on $D_{j(min)}$ which prevent the model from selecting the best possible allocation of resources.

Although it is not necessary to enter into detail, extensive experimentation with the model also indicates that useful improvements in equity (as indicated by the scatter procedure) can be obtained with a relatively little rearrangement of resources simply by reducing the permitted degree of change in each place of treatment. This is important, because unlike the simple example discussed above, health authorities will generally not be in a financial position to implement such large-scale changes in the short term. Given this situation, the approach can be used to investigate the case for incremental change by choosing the constraints accordingly. More details on how the model can be used in a planning context are given in Mayhew and Leonardi (1982).

To conclude, it should be added that the approach outlined above can be

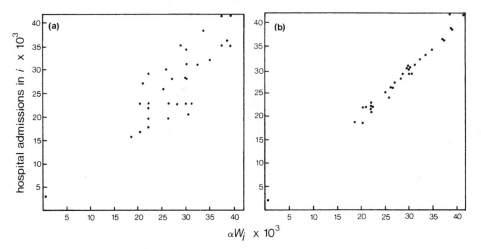

Figure 4.9 Scattergrams showing the relationship between hospital cases and the relative demand for hospital treatment: (a) on the basis of the existing spatial allocation of resources and (b) on the basis of the equity criterion.

extended and re-adapted in several different ways. Apart from examining policy objectives other than equity, it can be used to analyse the spatial provision of several different services in which allocations are dependent on the priority of each service. This refinement leads naturally to the question of building multilevel planning models of health service provision (Mayhew & Leonardi 1984). Once this stage has been reached it is clear that most of the mechanisms and concepts of central place theory may be made available in the form of a planning model. This means that the user can investigate many different scenarios involving assumptions about supply and demand without experiencing some of the inflexibilities of central place theory.

4.9   Conclusions

This chapter has been concerned with the problems of hospital provision in relation to the demand for health care services. It was argued that the demand for health care was difficult to separate from the need for health care and, historically, this had led to substantial differences in the levels of provision and patterns of utilisation of hospital services in different health care systems. It was further argued that the behaviour of patients in selecting particular hospital facilities was partly related to the problem of accessibility and partly to other factors, although under certain circumstances clear spatial patterns of demand within cities could be expected. Two hypotheses concerning the behaviour of demand were analysed to ascertain the implications for hospital services when there were changes in supply or demand. An analysis of data based on London supported the case for adopting the elastic demand hypothesis. A planning model incorporating this hypothesis was outlined and linked the earlier discussion. Finally, the planning model was reformulated so as to provide a practical tool that could investigate ways of promoting a more equitable distribution of resources. The reformulated model took into account the availability of resources, the accessibility of the population to supply, and various planning constraints within the health care system. The results showed that, in the case of London, a substantial redistribution of resources would be needed in order to achieve equity. This is not to argue that health authorities should follow this course exactly, even if it were possible to do so. The essence of the problem is to redistribute resources in a way which safeguards other objectives and takes into account constraints outside the health authorities' control. This can be achieved through the constraint mechanisms in the model. In the next chapter we look at the evolution of the London hospital system from an historical perspective, and show other aspects of the spatial imbalances highlighted in this chapter.

Appendix to Chapter 4

The first section of this appendix briefly summarises the mathematical details of the demand-led model to complement similar details given to the main chapter for the supply-led model. The second and larger section elaborates in more detail the reformulated model for allocating resources on the basis of equal access to health services for those equally at risk. For details of allocation models with formulations having other objectives see Mayhew and Leonardi (1982).

B1   *The demand-led model*
The demand-led model is written

$$T_{ij} = A_i W_i D_j f(\beta, c_{ij}) \tag{B1.1}$$

where T_{ij} is the flow of patients between area of residence i and place of treatment j in a group of acute hospital services; D_j is the attractiveness of hospitals in place of treatment j where attractiveness could be the number of available beds; W_i is the known demand in place of residence i for the same group of acute services; $f(\beta, c_{ij})$ is the deterrence function where c_{ij} are the accessibility costs of getting from i to j and β is the model parameter (this is later abbreviated to f_{ij}); A_i is the constraint variable which ensures the number of patients in i receiving treatment does not exceed the demand in i; that is,

$$A_i = [\sum_j D_j f(\beta, c_{ij})]^{-1} \tag{B1.2}$$

which ensures that

$$\sum_j T_{ij} = W_i \qquad (B1.3)$$

B2 The reformulated supply-led model

The objective of equalising access for equal risk of hospitalisation in the supply-led case (Eqn. 4.1) is equivalent to the minimisation of the following function:

$$\min_{D_j} \sum_i \left(\sum_j D_j B_j f_{ij} - \alpha \right)^2 = Z \qquad (B2.1)$$

where

$$\alpha = Q / \sum_i W_i \qquad (B2.2)$$

In addition, we want to ensure that allocations lie within the following limits:

$$D_{j(\min)} \leq D_j \leq D_{j(\max)} \qquad (B2.3)$$

and are subject to

$$\sum_{j \in L} D_j = Q \qquad (B2.4)$$

where L is the region of interest. By putting

$$B_j f_{ij} = \gamma_{ij} \qquad (B2.5)$$

expanding Equation (B2.1) and ignoring the constant term $I\alpha^2$, where I is the number of origins, we obtain

$$Z = \tfrac{1}{2} \mathbf{D}^\mathsf{T} \mathbf{A} \mathbf{D} - \mathbf{b}^\mathsf{T} \mathbf{D} \qquad (B2.6)$$

where $\mathbf{D}^\mathsf{T}$ is the transpose vector of resources $\mathbf{D}$,

$$\mathbf{D}^\mathsf{T} = [D_1, \ldots, D_j, \ldots, D_n] \qquad (B2.7)$$

A is a symmetric matrix composed of the following elements:

$$A = \begin{bmatrix} 2\sum_i \gamma_{i1}^2 & 2\sum_i \gamma_{i1}\gamma_{i2} & \cdots & & \cdots & 2\sum_i \gamma_{i1}\gamma_{in} \\ 2\sum_i \gamma_{i2}\gamma_{i1} & 2\sum_i \gamma_{i2}^2 & \cdots & & \cdots & 2\sum_i \gamma_{i2}\gamma_{in} \\ \vdots & \vdots & \vdots & & \vdots & \\ 2\sum_i \gamma_{ij}\gamma_{i1} & 2\sum_i \gamma_{ij}\gamma_{i2} & \cdots & 2\sum_i \gamma_{ij}^2 & \cdots & 2\sum_i \gamma_{ij}\gamma_{in} \\ \vdots & \vdots & & \vdots & & \vdots \\ 2\sum_i \gamma_{in}\gamma_{i1} & 2\sum_i \gamma_{in}\gamma_{i2} & \cdots & & \cdots & 2\sum_i \gamma_{in}^2 \end{bmatrix} = \{a_{ij}\} \quad \text{(B2.8)}$$

$\mathbf{b}^\mathrm{T}$ is the transpose of the vector $\mathbf{b}$ in which the elements are

$$\mathbf{b} = \begin{bmatrix} 2\alpha\sum_i \gamma_{i1} \\ 2\alpha\sum_i \gamma_{i2} \\ \vdots \\ 2\alpha\sum_i \gamma_{ij} \\ \vdots \\ 2\alpha\sum_i \gamma_{in} \end{bmatrix} = \{b_j\} \quad \text{(B2.9)}$$

Similarly (B2.3) and (B2.4) can be written in matrix notation:

$$\mathbf{D}_{(\min)} \leqslant \mathbf{D} \leqslant \mathbf{D}_{(\max)} \quad \text{(B2.10)}$$

and

$$\mathbf{C}^\mathrm{T}\mathbf{D} = Q$$

where $\mathbf{C}^\mathrm{T}$ is a $1 \times n$ vector transpose with all the elements set equal to one. Equations (B2.1), (B2.3) and (B2.4) have now been put into the standard form expected by a general quadratic programming algorithm. The matrix A is always positive definite or semi-definite, which indicates that global minima are obtainable. In an unconstrained problem the minimum of Z is found when the vector of first derivatives disappears; that is, when

$$g = \nabla(\tfrac{1}{2}\,\mathbf{D}^\mathrm{T}\mathbf{A}\mathbf{D} - \mathbf{b}^\mathrm{T}\mathbf{D}) \quad \text{(B2.11)}$$

$$= \mathbf{A}\mathbf{D} - \mathbf{b} = 0 \quad \text{(B2.12)}$$

Details of the solution method for this problem with and without constraints are contained in Fletcher (1971).

B3　The inferred deterrence function

An important assumption of the supply-led model in Equation (4.1) is that the β parameter is unchanged during the prediction period. This implies that for a given hospital and two arbitrary areas of residence k and i, the following conditions will apply

$$\frac{f_{ij}}{f_{kj}} = \frac{T_{ij}W_k}{T_{kj}W_i} \qquad (B3.1)$$

Providing that there are no major changes in transport costs or other factors affecting the values of f_{ij} and f_{ik}, the left-hand side of (B3.1) can be regarded as constant, further implying that

$$\frac{T_{ij}^p W_k^p}{T_{kj}^p W_i^p} = \frac{T_{ij}^o W_k^o}{T_{kj}^o W_i^o} \qquad (B3.2)$$

where the superscripts p and o indicate the predicted and observed variables. Re-arranging Equation (B3.2) we have

$$T_{ij}^p = q_{ij} p_{kj} T_{kj}^p \qquad (B3.3)$$

where

$$q_{ij} = \frac{T_{ij}^o W_i^p}{W_i^o} \qquad (B3.4)$$

$$p_{kj} = \frac{W_k^o}{T_{kj}^o W_k^p} \qquad (B3.5)$$

Letting $\sum_i T_{ij}^p = D_j^p$, the case-load of hospital j, we then obtain,

$$T_{kj}^p = \frac{D_j^p}{p_{kj} \sum_i q_{ij}} \qquad (B3.6)$$

so that

$$T_{ij}^p = \frac{q_{ij} D_j^p}{\sum_i q_{ij}} \qquad (B3.7)$$

or

$$T_{ij}^p = \frac{D_j^p W_i^p \; (T_{ij}^o / W_i^o)}{\sum_i W_i^p \; (T_{ij}^o / W_i^o)}$$ (B3.8)

In Equation (B3.8), the ratio T_{ij}^o / W_i^o plays the same role as f_{ij} in the original model, Equation (4.1) in the main text. For this reason and because it is based on observed data, this ratio is known as the inferred deterrence function. Further details are given in Mayhew *et al*. (1985).

5 *London's population and hospitals: 1801–1971*

5.1 *Introduction*

London forms a particularly appropriate case study on which to test the theory developed in the previous chapters, since it has strong traditions in hospital and medical work spanning several centuries, and the historical and statistical record regarding the establishment, sizes, locations and functions of the hospital facilities are long and reasonably accurate. The idea then is to examine the way in which London's hospital facilities have evolved in time and space, to identify trends in the locational behaviour between different types of hospital and to see where they differ, and to determine the extent to which an efficient and equitable pattern of services is being provided. The analysis is carried out in two parts; the first considers the evolution of the population of London at a general level of detail, and the second analyses the locational arrangements of hospital facilities at different dates. The period specifically considered is from 1801 to 1971, a time during which the hospital system and population of the city grew rapidly. Particular emphasis, however, is placed on the last 70 years of development, because the quality of the data is much better.

In evolutionary terms, the growth and distribution of London's population has followed a pattern subsequently reflected by many cities. At the beginning of the 19th century London was relatively small, but during the next 50 years it experienced a rapid influx and growth of population. Severe overcrowding in central areas of the city, widespread poverty and periodic epidemics were important features of this period and these were factors in subsequent growth in the hospital system. Towards the end of the 19th century residential development, facilitated by the growth of the railways, diverted some of the pressure on housing and the city began to grow rapidly in area. Further waves of building added successive rings of residential development to the perimeter of the city in the early 20th century, leading to a substantial deconcentration of population and further dispersal. In conjunction with these developments, widespread improvements were made to the public transportation system, facilitating longer journeys to work, for shopping and for recreation. These changes, however, were subsequently reinforced, and later overshadowed, by the rapid rise in car-ownership in the post-1945 period. This and related developments led to a considerable change in the mobility of the population, which in turn substantially modified the type of health care system that would be most appropriate for a more affluent and dispersed style of living. This does not imply, however, that every sector of society was equally affected by these trends. Shifts in population are highly selective and particular groups, such as the very old, poor or recent immigrants still tend to rely on residual services that have been allowed to run down in the wake of falling population levels. This then, in broad terms, is the situation today – namely, an essentially dispersed and very large city with an inner core of relatively deprived areas.

As will be seen, the spatial and temporal behaviour of the hospital system in response to these changes has been unusual in some respects and highly rational in others. Generally speaking, the growth in the hospital system outstripped the growth in population in relative terms, although certain types of hospital entered into and completed their growth cycles faster than others. One of the most important findings is that some hospitals tend to be spatially more mobile and responsive to population change than others. For instance, the locational pattern of psychiatric hospitals has remained relatively undisturbed for over 100 years as compared, say, with the locational pattern of acute hospitals.

In more detail, therefore, the task of this chapter is to establish the main features of the distribution of population and hospitals over the period in question, to analyse the locational patterns of different types of hospitals in relation to each other at different dates, and to establish the key points of similarity between the locational pattern of type 1 hospitals and the districting theory of hospitals developed in Chapter 2. The discussion starts with an analysis of population trends in the context of a growing urban area, and then considers data on hospital size and location. Selected aspects of these data are then analysed in more depth.

5.2   Characterising population densities in cities

The first problem is to provide a description of the urban population and area. If this description is too detailed, the aim of providing simple generalisations, applicable to other cities is made more difficult than if only the broad trends in population and area are evaluated. There are, of course, dangers in providing a too simplistic account of change, particularly at the very local geographical scale, and for this reason the analysis concentrates mainly on wider geographical issues. Perhaps the least acceptable simplification in the proposed approach is that it takes no account of the differences in the distribution of population by age and sex, which are known to be important factors in the utilisation of health services. However, these and other issues are readily able to be handled within the framework, but their consideration is deferred to separate later work which builds on this theoretical foundation.

A final question relating to the proposed level of detail is whether to consider the distribution of population as determined by census enumeration, or as it is actually distributed for a large part of each day in shopping centres, schools, offices, factories and so forth. Apart from the difficulties of measurement, the latter considerations are fortunately more of theoretical interest than of practical importance, the majority of patients using hospital services near their homes. In the case of accidents or emergencies, the situation differs and, as was seen in Chapter 4, a significant redistribution of risks is associated with a daily shifting population. Nevertheless, hospital admissions generated on this basis can be considered small as compared with normal admissions through alternative, more customary channels. Hence, the census or night-time population will be used.

The most convenient way to characterise the population of a city is with the aid of a population density function. Apart from concurring with the theory presented in earlier chapters (particularly Ch. 2), a descriptive model based on a suitable density function is consistent with the intended approach. The critical question is the choice of function and the adequacy with which it portrays essential features of the population of the city. In a wide and growing collection of studies, there is considerable agreement that the distribution of population densities in cities follows a negative exponential distribution. Clark (1951), Muth (1961), Mills (1970) and Bussiere (1972) are representative of early work in this area and in recent years this list has considerably expanded. Most of the possible alternatives to the negative exponential are in fact close mathematical relatives, but the empirical evidence for selecting one rather than another tends in fact to be fairly weak. Zielinski (1980) gives an abridged chronology of examples as indications of the range of variation that has been considered.

In choosing the negative exponential distribution from this list (or for that matter several of the others), it is essential to be clear on two specific assumptions – those of radial symmetry of the population density, and of monotonicity of the decline in density from the centre. Both are strong assumptions and consequently they restrict the possible inferences about the population that can be made. Essentially, the difficulty with the first assumption is that the city centre is actually relatively void of population (because commerce takes precedence over homes), but the residential density at this point is predicted by the density function to be a maximum. Similarly, suburban areas display peaks and troughs in residential density, corresponding to local centres and concentrations of population, whereas the second assumption predicts a smooth decline from the centre outwards. The implications stemming from both drawbacks are minimised, however, provided that the analysis concentrates on the *average* changes in density and hospital provision at different distances from the centre rather than on localised effects. In this mode of use, as will be seen, the approach gives an adequate description of the broad changes that have taken place. Nevertheless, the conclusions that can be drawn have to be qualified accordingly.

The negative exponential distribution

The purpose of this section is formally to introduce the negative exponential distribution and some of its key characteristics in this application. Mathematically, it is written as follows:

$$D(r) = Ae^{-br} \tag{5.1}$$

where $D(r)$ is the population density at distance r from the centre and where A and b are parameters. The first parameter is a measure of the extrapolated central density, as opposed to the actual density (because of the 'commerce' problem noted above), and the second is the exponential rate of density decline. From this basic equation, some useful measures can be obtained. For example, the population contained within a radius r of the city centre is

$$P(r) = 2\pi \int_0^r D(r)r \, dr = 2\pi A \int_0^r e^{-br}r \, dr \tag{5.2}$$

$$= (2\pi A/b^2)[1 - (1 + br) \, e^{-br}] \tag{5.3}$$

As r tends to infinity, Equation (5.3) becomes

$$2\pi A/b^2 = P \tag{5.4}$$

giving P the total population. Importantly, Equation (5.1) gives no natural value of r with which to bound the urban region and hence to determine the area. However, taking logarithms Equation (5.1) becomes

$$\ln D(r) = \ln A - br \tag{5.5}$$

Note that $\ln D(r)$ is zero when $D(r) = 1$, so that the radius r_{max} of the city may be defined from Equation (5.5) as follows:

$$r_{max} = (\ln A)/b \tag{5.6}$$

Clearly, r_{max} is still arbitrary, since it depends on the units of A and b (see also Table 5.2 below for examples), so that it is important to keep these units consistent. From Equation (5.6) the urban area may then be easily calculated, namely

$$S = \pi \left(\frac{\ln A}{b} \right)^2 \tag{5.7}$$

Consider now Equation (5.3) and let $\psi(r)$ be the slope of $P(r)$. Differentiating gives

$$\psi(r) = dP(r)/dr = 2\pi Are^{-br} \tag{5.8}$$

When $\psi(r) = 0$, a maximum value of the population is found on a ring whose radius is

$$r = 1/b \tag{5.9}$$

Now, dividing Equation (5.8) by the population P, determined in Equation (5.4) above, normalises $\psi(r)$ as follows:

$$\psi(r)/P = \sigma(r) = b^2 r e^{-br} \tag{5.10}$$

where

$$\int_0^\infty \sigma(r)\, dr = 1 \tag{5.11}$$

Whence, the mean distance $\bar{r}$ at which the population is located is readily calculated as

$$\bar{r} = \int_0^\infty \sigma(r) r\, dr \tag{5.12}$$

$$= \int_0^\infty b^2 r^2 e^{-br}\, dr = 2/b \tag{5.13}$$

The above equations may be used to analyse the distributions of populations in cities in several ways. Before considering this aspect, it is first necessary to consider how appropriate values for A and b in the density function for London may be estimated from population data and other studies.

A note on parameter estimation and choice of units

There are numerous cautionary points to note in fitting density functions to cities. The results are sensitive to the method of estimation, the areas of the population accounting units, and other factors (see, for example, Bussiere & Stovall 1981). In a very useful study for present purposes, Clark (1951) gave estimates for the parameters A and b in Equation (5.1) for London at five dates: 1801, 1841, 1901, 1921 and 1939. In order to bring this parameter set up to date, but also to be consistent with his results and therefore minimise possible problems of comparability, similarly derived estimates were made for 1951 and 1971, based on the linear regression of $\ln D(r)$ on r in Equation (5.5). One difference, however, was that Clark's estimated values for A and b in 1939 had to be revised for statistical reasons.

Clark measured population density in thousands per square mile and miles from the city centre. As noted above, however, the choice of units affects the size of the urban area as delineated by the circle of radius r_{max} defined in Equation (5.6). In determining which units to use, therefore, account was taken of the approximate extent of the built-up area of the city and whether it

Table 5.1 The effect of changes in measurement units on the area of London in 1801.

No.	Units A	b	A	b	P ($\times10^3$)	P' ($\times10^3$)	r_{max}	S (km²)
1	10^3 persons mile^{-2}	mile^{-1}	290.0	1.35	999.79	976.80	4.20 miles	143.53
2	10^3 persons km^{-2}	km^{-1}	111.969	0.8388	999.91	948.67	5.62 km	99.23
3	10^0 persons km^{-2}	km^{-1}	111969.2	0.8388	999.91	999.80	13.86 km	603.50
4	10^0 persons ha^{-1}	km^{-1}	1119.69	0.8388	999.91	992.75	8.37 km	220.09

could be comfortably enclosed within the estimated urban radius. An examination of different maps at varying dates indicated that units based on persons per hectare would be most suited to this purpose. The units of parameter A are therefore expressed in this form. The units of b, on the other hand, may be obtained by consideration of the dimensionality of Equation (5.4), the expression for the urban population:

$$\text{persons, } P = \frac{\text{persons/area}}{b^2} \quad \left(\text{i.e. } \frac{2\pi A}{b^2}\right) \tag{5.14}$$

This implies that

$$b^2 = [\text{length}]^{-2} \tag{5.15}$$

and that

$$b = [\text{length}]^{-1} \tag{5.16}$$

Because of the method of calculation there are further differences, also depending on the units, between the population P (as calculated in the model from $r = 0$ to $r = \infty$) and the population P' contained within r_{max}. In Table 5.1 values are given for P, P' and r_{max} in 1801 using four different sets of units. It is seen that in each case P' is quite close to P and, in fact, relatively insensitive in changes to r_{max}. This is simply because most of the population is clustered around the city centre and is thus relatively unaffected by the various estimates of the urban radius. A comparison of the estimated value of P' with the official population at this time also indicates little difference – 992 750 versus 959 000 – so that in this respect the density function seems accurate. In general, however, a poor correspondence between the official population and the estimated population is to be expected. This is because official figures are based on the population residing within fixed administrative boundaries, and these boundaries are usually very slow (certainly for London) in adjusting to population spill-over. As far as the present study is concerned, however, this

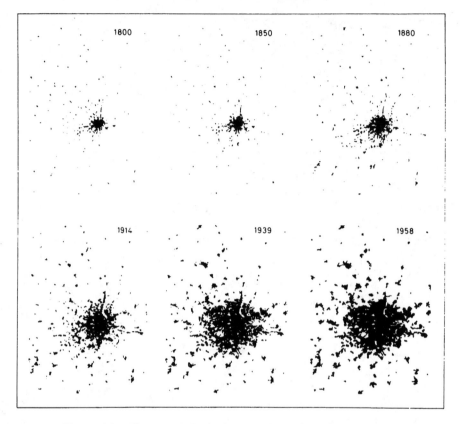

Figure 5.1   Changes in the built-up outline of London: 1800–1958.
Source: Clayton 1964.

spill-over defines the *de facto* size of the city region. Thus, the main theoretical advantage of using a density function is that the phenomenon of urban spread is automatically taken into account. Figure 5.1, taken from Clayton (1964) showing the changes in outline of the built-up London region, illustrates this particular argument. At the same time, however, it also focuses on another weakness of the density function approach. This is the tendency for the built-up area to fragment at the edges of the urban area. Again, as long as the discussion is confined to the general properties of the relationships between population density and hospital provision, it is assumed that such localised effects can be overlooked. Accordingly, attention now turns to an analysis of the spatial evolution of London's population. The main issues of concern are the changes in the distribution of population within the city and the urban area, and the identification of particular trends which have implications for the location of hospital services.

5.3   Interpreting the London parameter estimates

Parameter estimates based on the negative exponential density function using the methods described in the previous section are set out in Table 5.2. It shows the values of A and b at each date selected, together with estimates of r_{max}, the urban radius, and S, the urban area. The same results are also presented in graphical form in Figure 5.2, which is a plot of the logarithm of the expected value of the density (i.e. predicted by the regression estimates) on distance r for all the dates considered. A comparison of the results with the graph shows that there was a steady increase in the size of the urban area from 220 km² in 1801 to 7121 km² in 1971. After 1841, the value of the extrapolated central density, A, fell from 3089 to 193 persons ha⁻¹. This corresponded with a steady increase in

Table 5.2   Parameter values for London based on the negative exponential density function.

Date	A	$\ln A$	b^*	r_{max}	S (km²)
1801	1119.7	7.0208	0.84	8.37	220.1
1841	3088.8	8.0355	0.87	9.24	268.2
1871	1119.7	7.0208	0.40	17.39	950.1
1901	810.8	6.6980	0.28	23.96	1803.5
1921	695.0	6.5439	0.22	30.09	2844.4
1939	378.4	5.9360	0.16	38.22	4589.1
1951	263.5	5.5741	0.14	40.81	5232.2
1971	192.6	5.2606	0.11	47.61	7121.1

*By rounding b to two decimal places there may be small discrepancies if r_{max} is calculated using $\ln A/b$.

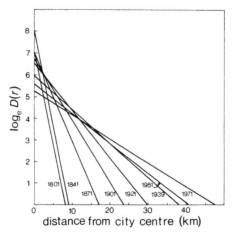

Figure 5.2   The relationship between the logarithm of population density (in persons per hectare) and distance (in kilometres) from the centre of London: 1801–1971.

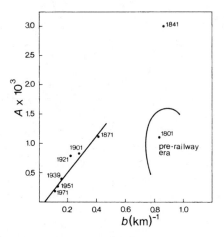

Figure 5.3   The parametric relationship between A and b, parameters in the negative exponential density function, from 1871 to 1971.

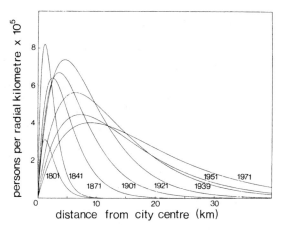

Figure 5.4   Inhabitants of London per radial kilometre on distance from the city centre: 1801–1971.

the peripheral population density, denoting the progressive suburbanisation of the city. Most interestingly, the value of b, the exponential rate of density decline, also decreased in value after 1841, indicating a lowering in average density in the subsequent period. Before 1841, however, residential densities were still increasing in all parts of the city. The reason why the reduction in average density appeared to commence after 1841 and not earlier can be appreciated if a graph is drawn of the respective values of A and b. This is shown in Figure 5.3 and, most importantly, it is a relationship common to many cities, particularly in developed countries (see, for example, Bussiere &

Table 5.3 Population and other statistics: 1801–1971.

Date	$1/b$ (km)	$2/b$ (km)	P' ($\times 10^6$)	P ($\times 10^6$)
1801	1.19	2.38	1.00	1.00
1841	1.15	2.30	2.56	2.56
1871	2.48	4.96	4.28	4.32
1901	3.58	7.16	6.46	6.52
1921	4.60	9.20	9.13	9.23
1939	6.44	12.88	9.68	9.86
1951	7.33	14.66	8.66	8.89
1971	9.05	18.10	9.64	9.91

Stovall 1981, pp. 151–66). The interesting feature in the case of London is that the relationship did not start to form until after the advent of the railway era.

Re-evaluating the population density profiles from a different standpoint, it is useful to compare further the results by constructing curves based on the function $\psi(r)$ in Equation (5.8). This is a measure of linear density and, as is seen in Figure 5.4, it portrays urban change as a succession of waves, first increasing in amplitude but then later decreasing. From the accompanying information in Table 5.3, it is seen that, after 1841, the peaks of the waves (which are located b^{-1} kilometres from the centre) began to migrate outwards. In the period from 1801 to 1971, the average distance at which the population was located increased from 2.30 to 18.10 km as a result. These movements, although accompanied by lower average settlement densities for the city as a whole (approximately 96 persons ha^{-1} in 1841 as compared with 14 persons ha^{-1} in 1971), were associated with large increases in the urban population. In 1971, for example, there was an estimated 9.64 million people, an almost fourfold difference as compared with 1841.

An acceleration of trends

The impression created by the results so far is of a steady, almost continuous, evolution in population densities and urban area; however, this is not entirely true. In London, as in other cities, more than one acceleration phase has been apparent during the course of urban development, with critical implications for the location of hospitals. In Figure 5.5, the population density at different distances from the city centre is plotted against the population of the city at each date. In particular, it shows a phase of rapid increase in inner-urban densities corresponding to high in-migration between 1801 and 1841; a phase of steep decline in inner-urban densities and an increase in intermediate densities, corresponding to the suburbanisation movement starting in the latter part of the 19th century; and finally, a second phase of steep decline in inner-urban densities, a fall in intermediate densities, and an increase in peripheral densities (>10 km), which correspond to a further spell of suburbanisation after 1920.

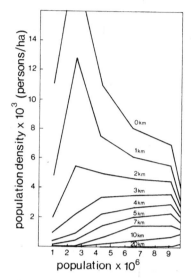

Figure 5.5   The density of population plotted against the total population of London at different dates and distances from the city centre.

Such changes as these clearly impact on the locational efficiency of hospitals in different ways. For example, one important issue, which was raised earlier in Chapter 4, is the extent to which the locational advantage of the centre has changed relative to other areas of the city. The centre has traditionally attracted many hospitals because it is the focus of a region containing millions of people and can be readily accessed by public transport. On the one hand an even larger population would reinforce its dominance, while on the other a much more dispersed population would support the opposite view. Is the centre, therefore, becoming more or less efficient? In fact it is possible to show using theoretical considerations (Mayhew 1979), that although the city centre *is* still locationally the most efficient of all locations in a city, its relative efficiency is greatly affected by the deconcentration of population. Figure 5.6, based on the theoretical analysis, shows the change in the average distance of the population in the London case from potential hospital locations at 0, 5, 8 and 10 km from the city centre. As is seen, the average after 1850 shows an increase for each point considered. From the non-centrally located hospitals, however, the increase is proportionately less, indicating that the relative differential has been somewhat reduced. Thus, in summarising the results so far, it has been shown that the population of London has evolved in such a way as to pose considerable problems for hospital authorities in terms of locating new hospitals and in deciding the fate of existing hospitals. The next phase of the case study is to evaluate London's hospitals within the above spatial population framework in preparation for a more detailed investigation of particular issues.

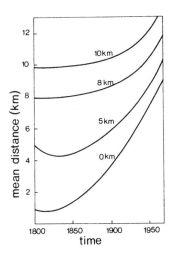

Figure 5.6 Changes in the average distance of London's population over time from hospitals located at 0, 5, 8 and 10 kilometres from the city centre.

5.4 The hospital stock

The first problem in collecting historical data is to decide on the definition of a hospital. This was particularly difficult in the case of the present study, because of the time period covered, the diverse historical sources that had to be employed, and the fact that the roles and functions of hospitals have tended to change, often imperceptibly, over time. At the same time it was necessary to ensure that the method of hospital classification would be reasonably efficient, particularly at dealing with types of hospitals that seemed to be borderline either between different categories of hospital or other types of institutions, such as nursing homes, clinics, health centres or dispensaries. There was also a need to establish the dates when hospitals either opened or closed, changed location to another part of the city, established annexes or branch hospitals, or expanded or contracted in size. It was also necessary to be aware of those hospitals, which though retaining the same location, changed their names in response to a change in the style of work carried out. A final complication was that throughout the period – at least up to the creation in 1948 of the National Health Service – the responsibilities for different categories of hospital were shared between different institutional bodies ranging from district councils and charities to, for instance, the Metropolitan Asylums Board which had responsibility for London psychiatric hospitals until 1929. All the different types, sources and variety of historical information that were used therefore necessitated careful evaluation, particularly with a view to ensuring consistency and comparability of information spanning different time periods. For convenience some of the background and details underlying the evaluation are contained in an appendix, which also provides a 'working definition' of a hospital based on a 1963 publication by the World Health Organisation. The latter proved to be a very useful guide for deciding whether to include certain marginal types of health care facility.

Locational maps

For the remainder of the chapter the principal aim is to present a preliminary analysis of several maps showing the locations of London hospitals over time. For reasons of poorer quality data, no detailed breakdown is supplied in the maps of hospitals for the period 1801–1870, as this would have been potentially misleading. However, a map is presented showing hospitals that were established during this period and which were still functioning at the beginning of the 20th century (information on hospitals that closed during the 19th century was not sufficiently reliable for mapping purposes). The focal point for all the maps is the centre of London – taken to be Charing Cross station – where the region covered has a radius of 50 km. Importantly, this radius is sufficiently large to incorporate all the preceding phases of urban development discussed earlier in the chapter. The specific contents of the maps are as follows:

Figure 5.7   Hospitals of all types
Figure 5.8   Type 1 hospitals treating patients in acute categories
Figure 5.9   Type 2 hospitals concerned with specialist care
Figure 5.10   Type 3 hospitals providing psychiatric care
Figure 5.11   Type 4 hospitals engaged in the longer term care and treatment
               of the old, disabled, and chronically ill
Figure 5.12   Type 5 hospitals treating infectious diseases
Figure 5.13   Hospital locations from 1801–1871.

The hospital types referred to in this list are the same as those discussed in the introductory chapter, in which the conceptual basis for this classification was presented. The precise technical specification used for defining each category are discussed further in the appendix to this chapter. At this stage, however, it is plain that there are a number of significant differences between the various locational arrangements.

As is seen, acute hospitals (Fig. 5.8) are the most numerous and have a distribution that is broadly aligned with the distribution of population. A close examination between the dates, however, shows that there has been a gradual thinning out of hospitals in the central area of the city and a consequent growth in the number of hospitals around the edge of the region. Because type 1 hospitals were argued to be most likely to conform to the requirements of central place theory, these sets of maps will form the core component of later investigations.

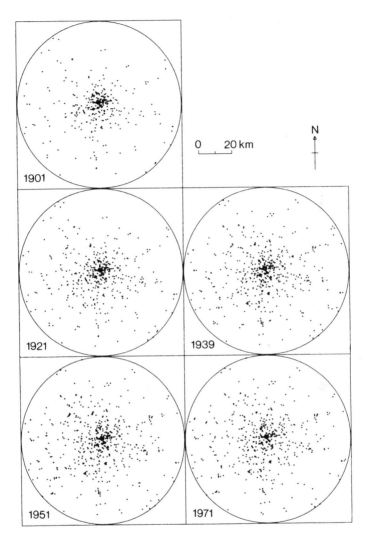

Figure 5.7 Locations of London hospitals: 1901–71.

Whereas the distribution type 1 hospitals appear, at first glance, to be related to population factors, the locations of type 2 hospitals (Fig. 5.9) are mainly concentrated in a small area in the centre of the urban region. Type 2 hospitals provide highly specialised health care services and attract patients from over a very wide area. It is commonly supposed that the focal area of these hospitals is associated with districts around Harley Street in London's West End, although in fact the locations are somewhat more widely dispersed than this. A critical factor in the location of specialist hospitals is plainly accessibility, and to this extent type 2 hospitals are related to the large, central teaching hospitals which,

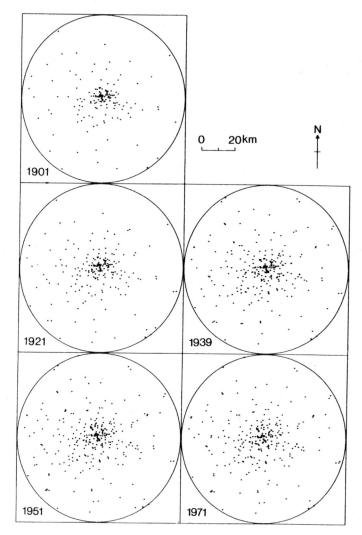

Figure 5.8 Locations of type 1 (acute) hospitals: 1901–71.

in addition to general services, also provide several of the same specialist services. The main distinctions are qualitative and relate to questions of prestige and tradition rather than to strictly defined catchment populations or hospital districts. To this extent, therefore, they compete with one another.

Type 3 hospitals for psychiatric care (Fig. 5.10) differ completely from either type 1 or type 2 hospitals in that they have a dispersed locational pattern, but with clumping in some areas. The largest psychiatric hospitals are very large indeed, and, measured in terms of the numbers of beds, they far outstrip the maximum size of any other type of hospital. The main psychiatric hospitals

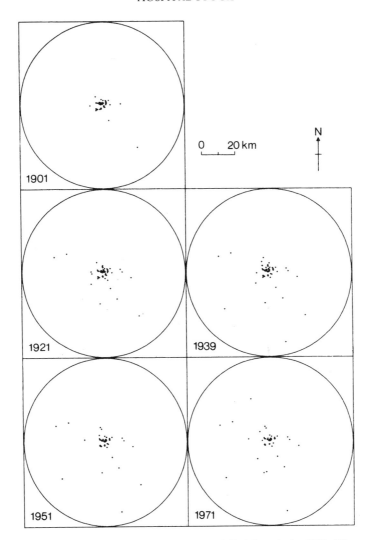

Figure 5.9 Locations of type 2 (specialist) hospitals: 1901–71.

are located in a ring between 15 and 30 km from the centre. When constructed in the 19th century, they were relatively inaccessible and removed from the main built-up area. In addition, the hospitals concerned were to a large extent self-contained institutions. The throughput of patients in such hospitals is relatively slow and most of the utilities, medical services and ancillary services are provided *in situ*. Although, as will be discussed below, there is evidence that the role and locational behaviour of type 3 hospitals is changing, their locational patterns over the period considered are clearly unrelated to the requirements of central place theory.

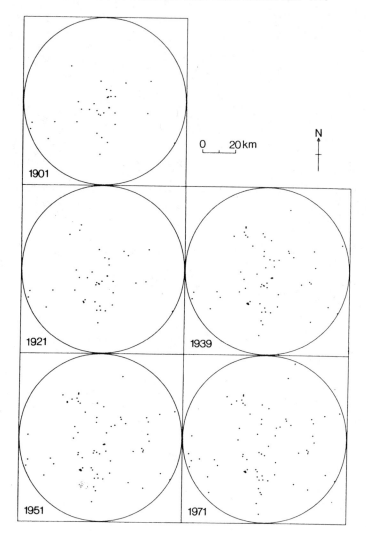

Figure 5.10   Locations of type 3 (psychiatric) hospitals: 1901–71.

Type 3 hospitals may be contrasted with type 4 hospitals (Fig. 5.11) which also have a similarly non-central locational behaviour. Unlike type 3 hospitals, however, type 4 hospitals exhibit considerably more flexibility in terms of their size, location and the suitability of different premises and other factors. Type 4 hospitals are concerned with the treatment and care of the chronically sick, elderly and disabled. Compared with other hospital types, this is an expanding sector of health care provision with, in the case of this city, a slight locational bias towards the western sector of the urban region. Most interestingly, a degree of the expansion is attributable to a change in function

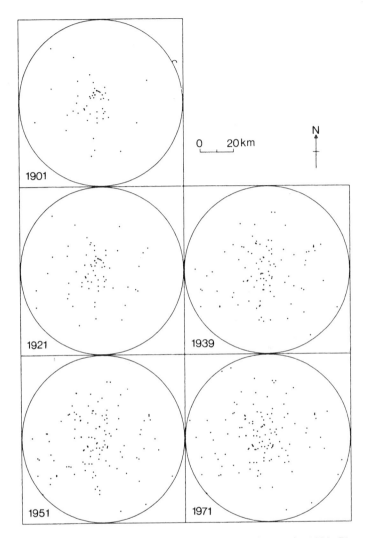

Figure 5.11 Locations of type 4 (long-stay) hospitals: 1901–71.

of some type 1 acute hospitals and this is a further aspect taken up later in discussion.

Type 5 hospitals, though from the maps very few in number today, provide an interesting locational category of their own (Fig. 5.12). These hospitals, which were used for the treatment and isolation of infectious diseases, have a statistically random distribution due partly to the nature of their function. The demand for the services of these facilities is intermittent, depending on the incidence of particular diseases. Many were constructed to deal with smallpox epidemics at the turn of the century and laws were enacted to ensure that local

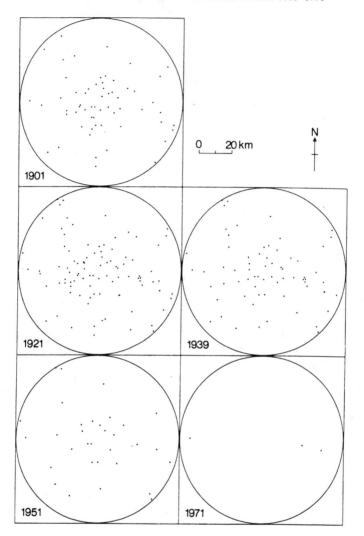

Figure 5.12   Locations of type 5 hospitals for infectious diseases: 1901–71.

administrative units would take some of the responsibility for provision. (In the earliest epidemics, the main form of isolation was provided by boats moored in the lower Thames, at Dartford.) Following the virtual cessation of serious smallpox outbreaks, the hospitals were used for the isolation of patients with scarlet fever and other infectious diseases (see also the appendix, Sec. C3). Due to improved methods of prevention, treatment, techniques of isolation and lower demand, the functions of these hospitals have now largely been absorbed by type 1 facilities, so that the majority of isolation hospitals are now closed. Not surprisingly, type 5 facilities were located mostly in isolated

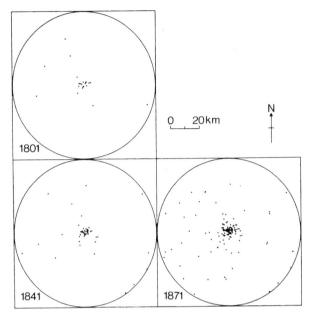

Figure 5.13   Locations of London hospitals: 1801–71.

places, although several very large 'fever' hospitals were located near the city centre, at strategic points of the compass, to cope with demand in the central area. These hospitals are now, however, largely engaged in other types of work.

The last preliminary point to make about the maps concerns Figure 5.13, showing the location of all hospitals at earlier dates. Although the question of historical evolution is examined in more detail later, it is probable that the pattern of expansion (from what is known of the history of this period) is slightly overstated. This is because, as has already been noted, information is omitted on these maps on any hospitals that closed before 1871. Nevertheless, there is a clear pattern in locational behaviour that is consistent with the earlier analysis of population trends.

The number and sizes of different hospitals
In this section the numbers and relative sizes of the hospitals are briefly compared with a view to isolating any important trends and to indicate whether or not there exist particular patterns that could have theoretical significance. In the discussion the measure of hospital size is based on the number of available beds. Although this method of measurement has certain disadvantages (see the appendix for details), its major value is its relative comparability at different times and the fact that the number of beds tends to be highly correlated with a broad range of hospital activities. Thus, for present purposes, it is taken as

Table 5.4 Basic hospital statistics: 1901–71.

Type of hospital	1901			1921			1939			1951			1971		
	N*	C̄†	%‡	N	C̄	%	N	C̄	%	N	C̄	%	N	C̄	%
Within 50 km															
1. Acute	152	155.7	44.8	196	179.4	43.1	260	206.5	47.4	277	215.5	50.7	257	214.6	50.5
2. Specialist	43	59.5	12.7	54	111.1	11.9	57	125.1	10.4	47	137.9	8.6	47	123.6	9.2
3. Psychiatric	33	809.7	9.7	41	1033.8	9.0	64	928.0	11.7	71	914.3	13.0	72	725.0	14.1
4. Long-stay	51	78.5	15.0	69	119.6	15.2	100	155.8	18.2	119	130.8	21.8	130	123.6	25.5
5. Infectious diseases	60	124.9	17.7	95	138.2	20.9	68	185.9	12.4	32	175.5	5.9	3	35.3	0.6
Total	339	190.1	100.0	445	230.6	100.0	549	270.3	100.0	546	278.9	100.0	509	254.1	100.0

* N = number of facilities within 50 km radius of centre.
† C̄ = average number of beds.
‡ Percentages may not total exactly 100 due to rounding.

being representative of the relative importance of hospitals within particular categories. Note, however, that due to the very different *nature* of the activities undertaken, this measure is less useful for comparing the relative importance of different types of hospital.

Table 5.4 provides a breakdown of the numbers and sizes of hospitals according to type and at different dates. Among other things it shows a decline in their number since 1939, in spite of an increase in the total population contained within the equivalent area. This decline is due mostly to the demise of type 5 hospitals already noted, but there has also been a slight decline in the number of type 1 hospitals. In terms of size, type 3 hospitals remain the largest, despite a decline in their average number of beds in recent years, exceeding the size of type 1 hospitals by a factor of three. Long stay and specialist hospitals are about equal in size, containing on average slightly over 100 beds.

From a theoretical standpoint, another interesting set of differences concerns the spread of hospital sizes around the average and the shift in this spread through time. Perhaps the most significant of these are the changes which have occurred in the case of type 1 hospitals. From a situation in 1901 of there being a few very large hospitals and many small hospitals, the position today indicates a significant decline in the numbers of the former and an increase in the latter. This shift in emphasis is indicative of a trend that, significantly, is absent from all the other types of hospital. Among other things, it is a trend that is closely related to changes in the distribution of population, and suggests that type 1 hospitals are particularly sensitive to this factor. The graphs in Figure 5.14 illustrate the change in the spread of hospital sizes in more detail. The size categories most affected by decline have been those with less than 100

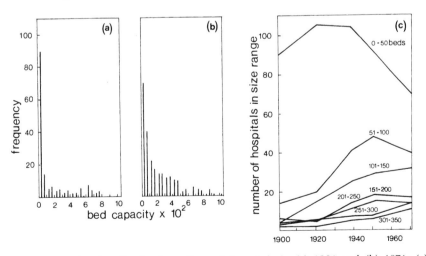

Figure 5.14   The size distribution of type 1 hospitals in (a) 1901 and (b) 1971; (c) shows changes in the frequency of different size categories.

beds; conversely, the number of hospitals with between 300 and 400 beds has
tended to increase. To a large extent, of course, this change has been effected
simply by adding extensions to smaller hospitals already in existence, although
plainly a large number of new hospitals have also been added during the
period. The size distribution of type 1 hospitals would thus appear to have
considerable significance in the context of the theory. On the one hand,
changes in the spread of hospital sizes would indicate possible changes in
districting patterns (since size and the area of a district are related), while on the
other, it would indicate possible shifts in the distribution of resources at
different levels within the administrative hierarchy.

Other considerations

To conclude the preliminary analysis, Table 5.5 provides details of the total
number of hospital beds by hospital type at different dates. Estimates of the
bed to population ratios are given, based on the values of the population P
given in Table 5.3. They also show important changes over time – changes
that are the outcome of a wide variety of social, medical and political factors.
The impact of these factors on hospital locations, however, is less obvious in
the sense that their effects are not easily attributed either to particular areas of
the city or to certain types of hospital. Notable exceptions, of course, are the

Table 5.5 The ratio of hospital beds to population levels; 1901–71

	1901	1921	1939	1951	1971
population, P' (×10⁶)	6.46	9.13	9.68	8.66	9.64
all beds	51 172	94 636	136 783	140 375	122 897
beds per thousand	7.92	10.37	14.13	16.21	12.75
type 1 beds	22 906	33 341	50 269	55 169	52 940
beds per thousand	3.55	3.65	5.19	6.37	5.49
type 2 beds	2 533	5 922	7 090	6 229	5 808
beds per thousand	0.39	0.65	0.73	0.72	0.60
type 3 beds	16 641	35 568	52 772	59 134	48 669
beds per thousand	2.58	3.90	5.45	6.83	5.05
type 4 beds	3 295	7 688	14 661	14 601	15 374
beds per thousand	0.51	0.84	1.51	1.69	1.59
type 5 beds	5 797	12 117	11 991	5 242	100
beds per thousand	0.90	1.33	1.24	0.60	0.01
independent/private beds	14 551	17 956	27 376	4 589	3 971
beds per thousand	2.24	1.97	2.83	0.53	0.41
public/state beds	36 621	76 680	109 407	135 786	118 926
beds per thousand	5.67	8.40	11.30	15.68	12.34

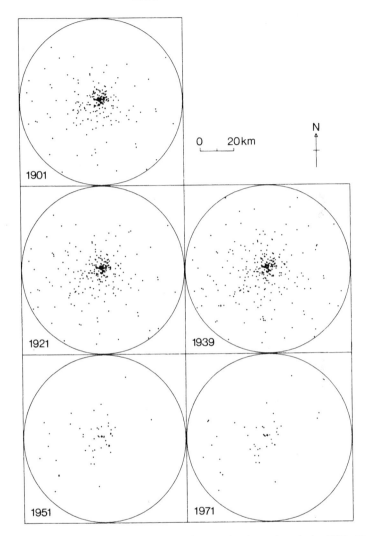

Figure 5.15 Locations of independent and private hospitals: 1901–71.

decline in type 5 hospitals (due to the reduction in infectious diseases) and the
increase in type 4 hospitals (due to the increase in the number of elderly). One
important perturbation in terms of the way hospitals in London are organised
that can be mentioned, however, arose out of the nationalisation of hospital
services in 1948. Hospitals, until then mostly independent, were offered the
choice of working within the newly created National Health Service or
continuing to operate separately as private institutions. Figures 5.15 and 5.16
show the subsequent change in locational configuration resulting from this
administrative reorganisation. Before 1948, hospitals are classified as either

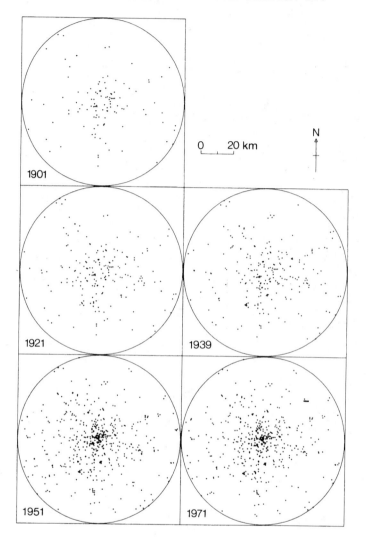

Figure 5.16 Locations of public and state-run hospitals: 1901–71.

'independent' or 'public' institutions; after 1948 they are re-termed either 'private' or 'state-run'. Although the total number of hospital facilities at this turning point in the system remained the same, it is plain that there was considerable geographic selectivity as regards which chose to join the nationalised health service and those which did not. The main reason for this was that most of the 'public' hospitals before 1948 were of types 3, 4 and 5, whereas type 1 hospitals were mainly independent. With the exception of a small private sector consisting of mostly type 1 facilities, predominantly in the west of the city, most type 1 hospitals opted to become 'state-run'; hence the changes seen.

5.5   Changes in the districting of London hospitals

The preliminary analysis of the hospital data is now complete and we turn next to some of the more detailed aspects of the locational patterns associated with type 1 hospitals. The first problem of interest is the nature of the districting pattern underlying their distribution, and whether it has changed as a consequence of a changing population and other factors. This is important because in the districting theory of hospitals presented in Chapter 2 different implications, in terms of either the equity or the efficiency of the hospital system, were seen to be contingent on the pattern of locations. In the London case, a good indication that the districting patterns have in fact changed is provided by Figure 5.17. This shows, in proportionate terms, the cumulative distributions of population, hospitals and hospital beds at different distances from the city centre in 1901 and 1981. Plainly, the results differ in a number of respects. Firstly, the urban radius r_{max} is much reduced in 1901 as compared with 1971; secondly, the relative shapes of the distributions differ.

To interpret the changes of shape, it is necessary to consider how the distributions should have behaved, if all the hospitals had been districted according, for example, either to the P-criterion or to the MC-criterion. The P-criterion, it will be recalled, locates hospitals so that each one serves the same population, whereas the MC-criterion locates hospitals so that no one is farther than a given distance from their nearest hospital. In terms of the cumulative distributions the P-criterion should result, therefore, in the individual

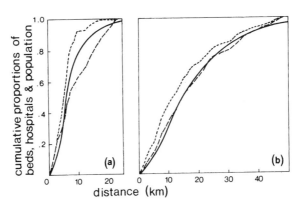

Figure 5.17   Cumulative distributions of type 1 hospitals and beds, compared with the distribution of population in (a) 1901 and (b) 1971. The radius of the city was 24 km in 1901 and 48 km in 1971 (see Table 5.2). ————, Population; – – – – –, beds; – – – – –, hospitals.

distributions for population and hospitals being superimposed because each area of the city would receive an equal proportionate allocation of hospitals. In the case of the MC-criterion the distribution would increase in proportion to the square of the distance from the city centre. This means that the distribution would be beneath the population curve and curve inwards instead of outwards. The cumulative distribution of beds, by contrast, should be the same in both cases as the distribution of population, since a condition of both criteria is that the number of beds allocated per unit area is in proportion to the population. In fact there will be slight differences because allowance has to be made for the hierarchical effect which entails provision for large hospitals located near the city centre. This would result in the theoretical distribution of beds rising more steeply at first.

If we consider again Figure 5.17, it is now possible to re-interpret the shift seen in the two sets of distributions. In 1901, the distribution of hospitals would indicate that there were relatively more hospitals than there were people in peripheral areas of the city. In 1971, by contrast, the distribution of population and hospitals were clearly in closer balance. It is noteworthy, however, that there is a relatively larger surplus of beds in the inner areas (less than 15 km from the centre, say) as compared with 1901. This is an important observation because it means that the distribution of beds has not kept entirely in step with the deconcentration of population.

5.6   Changes in the hierarchy of acute hospitals

An interesting set of changes accompanying the shifts in these distributions concerns the relative sizes of hospitals in different areas of the city. Figure 5.18 shows how the average size of hospital changes with distance from the city centre at five dates. The results are indicative of the adaptive nature of the system and the structural changes which occur. The following trends are worthy of note:

(a)   *1901*   The average size of hospital reaches a peak between 5 and 7 km from the centre. In peripheral areas the average falls dramatically to about 10 beds. This is because hospitals in these areas were mainly of the 'cottage' type catering only for the immediate locality.

(b)   *1921*   The inner peak remains, but it is accompanied by a second, smaller peak at about 25 km, denoting the emergence of a second hierarchical layer.

(c)   *1939*   The trend started in 1921 is continued, but the emergence of a third, intervening peak is apparent at 15 km.

(d)   *1951*   Trends commenced at previous dates are maintained, but there is a levelling in the amplitude of the three peaks shown.

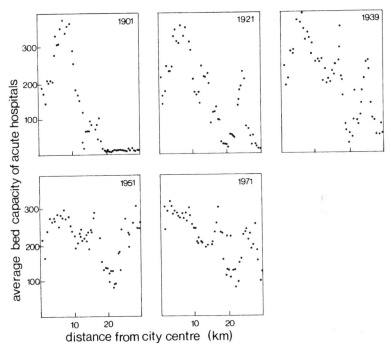

Figure 5.18 Variations in the average size of type 1 acute hospitals with distance from the city centre in London in 1901, 1921, 1939, 1951 and 1971.

(e) *1971* The inner peak has now moved closer to the city centre. This effect is explained by a further closure of smaller facilities raising the average size of hospitals remaining.

The timing of growth

The trends highlighted above may be broadly linked to the dates at which different facilities were established. Two useful pointers towards these linkages are the dates when, respectively, 10 per cent and 50 per cent of the eventual (i.e. 1971) stock of hospitals were positioned. These measures might be said to define two key stages in the growth cycle. In particular the 10 per cent level is an indicator of the 'take-off' point because it usually denotes a point in the evolution of the hospital system prior to a rapid increase in the rate of growth. The 50 per cent level, by contrast, denotes the halfway mark in the cycle, following which the rate of growth declines. Table 5.6 provides details of the results obtained. They are divided into three parts. Part (a) shows the dates for the 10th and 50th percentiles based on an analysis of the hospitals established in five rings, each 10 km in width, around the city centre. Part (b) shows equivalent dates but this time by hospital type. Note that type 5 hospitals are

Table 5.6 Take-off dates and mid-points in the growth of London hospitals.

(a)   Type 1 hospitals by ring

	Ring				
	1	2	3	4	5
10th percentile	1824	1883	1898	1859	1850
50th percentile	1887	1915	1918	1926	1922

(b)   By hospital

	Type				
	1	2	3	4	All
10th percentile	1843	1837	1863	1868	1850
50th percentile	1897	1889	1910	1912	1902

(c)   By administrative type

	Public	Independent
10th percentile	1859	1834
50th percentile	1904	1890

excluded because the number surviving to 1971 were too few to draw any conclusions. Part (c) shows the same results by administrative type of hospital (either public or independent). The terminal date for this section of the Table was 1947 (not 1971), reflecting the important point that after this time the distinction between public and independent hospitals changed (see also Sec. 5.5).

In analysing the results it should also be noted that the data include hospitals which, in the course of their operation, were converted from different types of hospital; in their cases the dates of origin were suitably modified. From part (a) it is seen that the take-off dates and mid-cycle dates change order slightly. Innermost hospitals grew fastest initially, followed by hospitals in the outer rings. During the intermediate phase hospitals in the middle rings overhauled those in the outer rings and this is reflected in the sequence of dates at the mid-point. In part (b) the results indicate that the equivalent stages in the growth cycles of type 3 and type 4 hospitals occurred very much later than for type 1 and type 2 hospitals. This result reflects the important point that type 3 and type 4 hospitals were, at the time of their construction, publicly financed and operated. They filled important gaps in the existing health care services which were not adequately covered by the independent sector. This conclusion is underlined in part (c) of Table 5.6, which shows that the public sector attained the 10th percentile 25 years later than the independent sector.

Changes in hospital function

In some cases, a change of hospital function provides an alternative to closure. However, because of the diverse locational requirements of different types of hospital (e.g. see Figs 5.8–5.12), the opportunities for functional change are strictly limited. The scope for change also depends, of course, on other factors, such as the suitability of the buildings. A useful working hypothesis, however, is that if the locations of one type of hospital are correlated with the locations of another then the chance of their exchanging functions should be larger than if there were no correlation. To test this hypothesis the London area was divided into a grid containing slightly more than 300 cells each 25 km² in area. The frequency of association at each date between hospital types in each cell was determined using a 2 × 2 matrix of association. Table 5.7 shows the general form of the matrix. A symbol r denotes the number of cells in which hospitals of type j are observed in the same cells as hospitals of type i. If N is the total number of cells altogether, R is the number occupied by hospitals of type j and n the number occupied by type i, then the remaining algebraic quantities in the cells, including the row and column totals, can be simply determined. In such tables more than one statistical test can be applied to test whether or not there is any association between the two variables, in this case two types of hospital (for examples and details, see Lewis 1977, Ch. 2). Intuitively, if there were a stong degree of spatial association between the hospitals concerned, the values in the top left and bottom right boxes in the matrix would dominate those in the bottom left and top right boxes.

In an exhaustive set of tests involving 50 pairs of potential associations on the data, the results showed conclusively that type 1 and type 4 hospitals had the strongest spatial links over time. Referral to Figures 5.8 and 5.11 shows the plausibility of this conclusion, where it is seen that these hospitals dominate the

Table 5.7 A 2 × 2 frequency table of geographical association.

		Hospital type (j)		
		+	−	
Hospital type (i)	+	r	$n - r$	n
	−	$R - r$	$N - R$ $-n + r$	$N - n$
		R	$N - R$	N

Table 5.8 Results obtained for the ϕ coefficient, showing the relative strength of spatial association between different hospital types in 1939.

1939 Type					
1	1.00				
2	0.22	1.00			
3	0.20	0.21	1.00		
4	0.35	0.29	0.28	1.00	
5	0.33	0.15	0.17	0.22	1.00
	1	2	3	4	5

i (row), j (column)

urban area both numerically and in terms of their geographical spread. At times when they were more numerous type 5 hospitals also had strong links with type 1 hospitals (but, interestingly, not with type 4 hospitals). Types 2 and 3, however, have the least spatially in common with other hospitals and, again, referral to Figures 5.9 and 5.10 provides intuitive confirmation of this.

Table 5.8 shows an illustrative set of results for 1939 based on one of the procedures used. Entries show the value of Kendall's ϕ coefficient, a measure of correlation, which varies from -1 (perfect dissociation) to $+1$ (perfect association). The ϕ coefficient is calculated from the following formula:

$$\phi = \frac{O_{11}O_{22} - O_{21}O_{12}}{\sqrt{[(O_{11} + O_{12})(O_{21} + O_{22})(O_{12} + O_{22})(O_{11} + O_{21})]}} \tag{5.17}$$

where O_{ij} is the frequency observed in box ij of the association matrix in Table 5.7.

To see whether the degree of spatial association is related to the propensity for hospitals to exchange functions, we turn to Table 5.9. This shows the actual percentage frequency of changes in hospital functions between 1901 and 1971 by hospital type. In all, 81 cases of functional change were detected among the data and, as is seen, the main movements are from type 5, type 1 or type 4 hospitals. Two other notable features are the changes from type 1 to type 4 hospitals and from type 4 to type 1 hospitals. It is noteworthy that type 3 hospitals are the least likely of all hospitals to change function and, indeed, from the analysis of spatial associations these hospitals are also the least correlated with the locations of other hospitals. Importantly, this suggests that, as a result of their locations, they are unsuited for conversion.

Table 5.9 The percentage frequency of changes in hospital functions between 1901 and 1971 by hospital type.

	Type	1	2	3	4	5	Row total
				to			
from	1	—	1.2	1.2	16.0	0	18.5
	2	4.9	—	1.2	0	0	6.2
	3	0	0	—	2.4	0	2.4
	4	8.6	2.4	1.2	—	0	12.3
	5	18.5	0	2.4	39.5	—	60.6
column total		32.1	3.7	6.2	58.0	0	100

The conclusion from this excerpt based on a much wider analysis has shown, therefore, that there is a broad connection between the propensity for hospitals with a similar locational behaviour to exchange functions. The changes are a reflection of a complex pattern of evolution and growth involving each sector of the health care system but are also partly driven by the changing location of population and demand.

5.7 Some details of the urban hierarchy: 1971

To conclude our analysis of London hospital locations, we turn now to a specific application of districting theory. An advantage of the theory is that, in principle, it enables a more systematic analysis of locational patterns in a city. As illustration of this, it is of interest to focus on the locational pattern of type 1 hospitals as it existed in 1971, and to consider whether, in certain points of detail, the spread of hospitals could have been improved. There are essentially three stages involved in the procedure: firstly, a decision is taken on the type of districting pattern desired and the number of levels in the proposed hierarchy; secondly, the existing stock of hospitals is relocated according to the theoretical pattern; thirdly, the theoretical pattern is compared with the actual pattern to see where and if there is a relative over-provision or under-provision of hospital facilities. A distinction should be drawn between the first and second steps, which are administrative decisions involving a prior consideration of many economic and social factors, and the second and third steps, which are based on the districting theory.

The London case
According to Table 5.2, the urban radius for 1981 was 47.61 km, within which were located 248 type 1 hospitals, nine less than the total contained within a 50 km radius (Table 5.4). The largest hospital, with a rank equal to 1, contained 1040 beds; the smallest, with a rank equal to 248, contained 12 beds.

Suppose now that it were decided to opt for a P-districted system. As Section 5.6 showed, the locational pattern in 1981 already closely matched the P-criterion and so, it is hoped, only marginal adjustments to the existing set of locations will be needed. Suppose also that a four-tier system provided the appropriate hierarchical structure. We recall now the discussion in Section 2.11, where the concept of the multiplier was discussed and a practical example, similar to that being considered here, was given. The main difference now is that a multiplier of 4 instead of 3 is selected. This means that if the number of hospitals in level one is one, subsequent levels will contain $(4 - 1)$, $4(4 - 1)$ and $4^2(4 - 1)$ hospitals respectively. This is equivalent to the sequence shown in Table 2.2, column 3, namely a, $a(l - 1)$, $al(l - 1)$ and $al^2(l - 1)$, where a is the number of hospitals in the first level and l is the multiplier. The required pattern, however, will not match this sequence exactly because some hospitals will be shared with the external, non-urban region. To proceed, therefore, we start by assuming a uniformly distributed population in a region of the same radius and then calculate the radius of the smallest district, assuming 248 hospitals altogether. If S is the size of the urban area, d (the radius of the smallest district in level 4) is given by

$$d = \left(\frac{2S}{248 \times 3\sqrt{3}}\right)^{1/2} \tag{5.18}$$

Here the formula for the area of a hexagon in Section 2.7 has been used. This information is sufficient geometrically to construct the required districting pattern. This pattern shows that there should be seven, 12, 42 and 187 hospitals in each of the four levels. This information is recorded in the rank-size distribution shown in Figure 5.19 in the form of the vertical lines, a, b and c. Reference to the vertical axis shows that the bed range allocated to each level is 800–1040, 651–759, 334–650 and 0–333 beds. Note that had we not taken into account those hospitals physically located inside the region which serve external areas, the numbers in each level predicted by the multiplier formula would have been 3.9, 11.6, 46.4 and 186.0. This would imply, for example, that only slightly over half the facilities provided by level 1 hospitals are actually designated for residents of the urban region itself.

Transforming the theoretical pattern
Thus far, we have determined the size range of the four levels and how many hospitals are contained in each level. The next problem is physically to locate the hospitals in the urban area so that they correspond with the P-criterion, given that the underlying distribution of population is not uniform but rather is broadly related to the negative exponential distribution. To achieve this, we draw on the method of transformations introduced in Section 2.8. Specifically, given a negative exponential distribution in 1971 with parameters shown

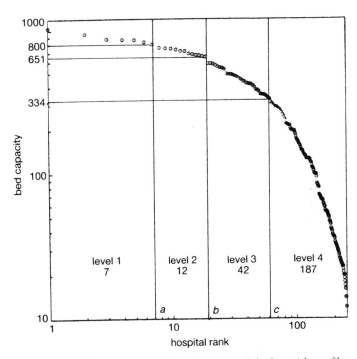

Figure 5.19   A plot of the logarithm of hospital size and the logarithm of hospital rank in 1971. Lines indicate the divisions into a four-level hierarchy (see text).

in Table 5.2, the new locations of the hospitals may then be determined in terms of their distance from the city centre. In the exercise, there is some indeterminancy as to the rotation of the theoretical pattern. This has to be determined on a trial and error basis. Here, to save on detail, we are concerned only with the broader aspects of comparison. In particular, suppose a circular area of uniform population density were divided into five rings each 10 km in width. Then further suppose the radii of the rings were transformed in accordance with the 1971 parameter estimates for London. The widths of the rings would be 3.01, 3.9, 5.54, 9.96 and 25.2 km respectively, which added together give 47.61 km – the radius of the region. We can now examine how many hospitals in each level are actually located in these rings as compared with the number that should be in them according to the theory.

Figure 5.20 shows the actual locations of hospitals and symbols are used to denote their levels in the hierarchy. Table 5.10 compares the actual number in each ring by level with the theoretical level and the observed difference. For example, in ring 3 level 2 the theory predicts six hospitals whereas the actual number located is only three, making a deficiency of three hospitals.

From Table 5.10 three important conclusions may be drawn with implications for the future location of hospital facilities in London. Firstly,

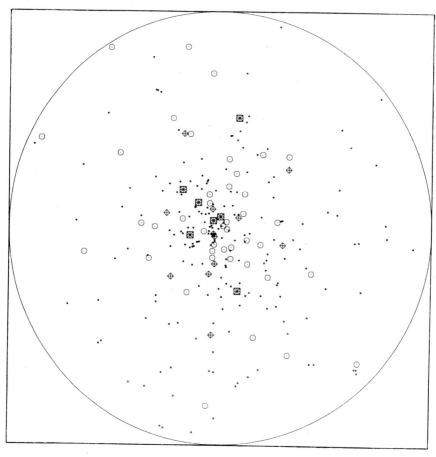

Key

⊞ > 800   ⬧ 651–800   ⊙ 334–650   ▫ 10–333 beds

Figure 5.20 The actual distribution of type 1 (acute) hospitals in London in 1971, showing the hierarchical levels; ■, Level 1; ◆, Level 2; ○, Level 3; ●, Level 4.

there is a relative *surplus* of 15 facilities in the first and second rings, two of the surplus being level 1 facilities. Secondly, there is a total deficit of 15 facilities in the last three rings, two in level 1, four in level 2, and 9 in level 4. Thirdly, in ring 3 the number of level 4 facilities, the smallest, is 13 in deficit. The results show, therefore, some structural imbalance in the locations of acute hospitals in several important points of detail. The problem of the deficit of level 4 facilities in ring 3 could be partially solved relatively easily by a downgrading of level 3 facilities, and there are further possibilities. The most difficult

Table 5.10 Actual and expected number of hospitals in each urban ring by level.

Ring	Level 1			Level 2			Level 3			Level 4		
	a^*	$b^†$	$c^‡$	a	b	c	a	b	c	a	b	c
1	2	1	1	2	0	2	3	0	3	14	12	2
2	1	0	1	2	0	2	9	12	−3	25	18	7
3	2	0	2	3	6	−3	7	0	7	35	48	−13
4	1	0	1	4	6	−2	11	18	−7	53	48	5
5	1	6	−5	1	0	1	12	12	0	60	61	−1
Total	7	7		12	12		42	42		187	187	

$^*a$ represents the actual number.
$^†b$ represents the expected number.
$^‡c$ represents the difference, $a - b$.

problem to solve, however, is the over-provision of level 1 hospitals in the inner rings. More provision needs to be made at the periphery of the region, but these hospitals, with many teaching and specialist functions, are very expensive to develop.

5.8 Conclusions

To arrive at this point, a considerable amount of theoretical and empirical ground has been covered. The objective has been to develop and apply some general principles explaining the locations of hospitals and hospital services in cities. In pursuing this objective it was realised, from the outset, that there would be many pitfalls, and this proved to be the case. Indeed, it would be surprising if, in analysing something as complicated as the organisation of health care, none was forthcoming. As a result it was seen that, in detail, the problem of locating hospitals in cities was highly complex, but by concentrating on selected aspects progress could be made in understanding the reasons for particular patterns of provision and for different configurations of hospital facilities in space and through time.

To recap briefly, the basis for the analysis was provided by central place theory, which was originally developed by Christaller to explain the spatial arrangement of cities, towns, villages and other settlements in a region. To apply this theory to hospital systems within cities, therefore, certain changes had to be made to the basis of the theory. Besides a reduction in the geographic scale of enquiry, these changes included new assumptions about the objectives of a central place system; the introduction of different forms of demand behaviour; alternatives to the standard, regular hexagonal set of market areas; different measures of accessibility costs; and some very simple dynamics to deal with the internal changes in the size and distribution of population in city regions.

A method of transformations was developed to enable the development and costing of idealised patterns of provision that could be evaluated against different assumptions concerning population density, demand behaviour and accessibility. The resultant systems of districts were seen to be relevant in different circumstances depending on the type of hospital facility and whether an equitable or efficient pattern of provision was desired. Together these analyses provided the framework for the case study based on the London hospital system, which included a spatial evaluation of accident and emergency centres, a spatial analysis of outpatient patterns, a presentation of a planning model for acute services which functioned at the health district level and, finally, in this chapter, an historical analysis of trends in the location of population and hospitals.

In broad terms, the results show conclusively that different approaches are needed for analysing different spatial problems. Much depends on geographical scale, in particular the extent of the geographical impact of different health care services. Given the finite quantity of resources available for health care, the study has indicated the potentially large contribution of location theory for providing a more equitable and efficient health care system in cities.

Appendix to Chapter 5: background sources and assumptions used in the case study

This appendix describes the background to the case study, the sources of data used, the classification scheme, assumptions underlying the data, and unavoidable omissions. The sections consider

C1 the definition of a hospital and the sources of information
C2 the dates of hospital origin
C3 the classification of hospitals
C4 the distinction between 'private' and 'public' hospitals
C5 the measurement of hospital location
C6 the measurement of hospital size
C7 summary hospital statistics.

C1 The definition of a hospital

Pinker (1966), in *English hospital statistics 1861–1938*, applied to historical sources the following definition of a hospital, as set out by the World Health Organisation (1963):

'. . . an establishment which offers accommodation and provides medical and nursing care to persons who are sick or injured or are suspected of being sick or injured, to women during childbirth, or to more than one of these categories. The term includes, therefore, not only institutions with beds for patients which are described as hospitals (both general and specialised) but also appropriate parts of hospices or homes providing geriatic care. It includes nursing homes, night sanatoria, but not rest homes, holiday homes and preventoria which accommodate patients from homes where there is a case of tuberculosis, but who have not themselves developed the primary type of tuberculosis.'

Where practical, this definition was also employed in the case study. It excludes day clinics, dispensaries, day hospitals, and health centres that do not offer accommodation. The chief problems in using it concerned peripheral institutions – nursing homes, etc. – whose precise functions, and hence hospital status, could not always be reliably established. Decisions for exclusion or inclusion rested mainly, therefore, with specialised directories, or with census evidence. *Burdett's hospitals and charities: the year book of philanthropy and hospital annual (1889—1930)* and the *Hospitals yearbook (1931—)* are the most important directories in the hospital field. The first, which changed its name in 1930 to the *Hospitals yearbook,* established high standards of comprehensiveness, organisation, and consistency. Initially at least, the excellence of Burdett was not maintained by its successor, since it failed to provide a full guide to hospitals in the public sector. This gap was filled by the publication in 1945 of the *Hospital survey,* a government census of hospital accommodation in 1938.

The sections of Burdett's directory that qualified for consideration were: London hospitals; London hospitals for infectious diseases; provincial hospitals (within 50 km of the centre of London); provincial Poor Law hospitals and infirmaries; provincial hospitals for infectious diseases; lunacy boards, mental hospitals, and asylums; convalescent homes, consumption sanatoria, institutions for chronic and incurable cases, and retreats for inebriety and abuse of drugs.

In the post-1948 period, following the creation of the National Health Service (NHS), the flow of hospital data further improved. Also, the *Hospitals yearbook* re-established its former comprehensiveness, and apart from a small number of unrecorded private hospitals most hospitals are now incorporated. The format of the directory in 1971 reflected the old administrative structure of the English health service before the 1973 administrative reorganisation. In terms of the case study, therefore, the sections of interest were the metropolitan regions, teaching hospitals, a portion of the Oxford region, and relevant hospitals in the private sector.

Particular care was required in dealing with Poor Law and other institutional information, especially that dating from before 1930. Many hospitals under the Poor Law formed only part of a larger institution dealing with the sick and able-bodied pauperised; some were also upgraded workhouses, while others were purpose-built (Abel-Smith 1964). Some difficulties were experienced in determining which of these institutions qualified under the definition of a hospital. Burdett only included establishments that had training facilities (Pinker 1966). Pinker noted, however, that without the contribution of the workhouse sector, an inaccurate impression of inpatient accommodation would be obtained. Additional sources of information that were found to be helpful in this respect included the following: *The LCC hospitals: a retrospect* (LCC 1949); Part II of *A joint survey of medical and surgical services in the administrative county of London – municipal hospitals, clinics, and dispensaries* (1932–1933); *A survey of hospital and institutional accommodation*, Vols I–III (LCC 1929); and *Records of the Boards of Guardians, Form A weekly returns* (1901, 1921; see under LCC in references).

For institutions dealing with mental defectives, useful sources included: *General survey of children's institutions and mental hospitals (Metropolitan Asylums Board), casual wards and miscellaneous buildings* (LCC 1929); *England's first state hospitals, 1867–1930* (Ayers 1971); *The Metropolitan Asylums Board and its work, 1867–1930* (Powell 1930); and *A list of state institutions, certified houses and approved homes for mental defectives* (England and Wales' Board of Control 1938). The Metropolitan Asylums Board was the central authority in London for providing and maintaining hospitals and institutions for infectious diseases, mental defectives and epileptics, the London poor and other sick and deprived groups. Under the Local Government Act (1929), up to the creation of the National Health Service, control passed to the local authorities; for London, this authority was the London County Council (LCC).

Nursing homes were rarely listed by the specialised directories. However, classified telephone directories, which began publication in the 1930s, gave names and addresses of numerous establishments describing themselves as nursing homes. If these homes indeed provided medical care, it would imply the existence of a large tier of health care provision unaccounted for in all the major data sources (including the *Hospital survey*). It had to be assumed, therefore, that this group of institutions was relatively unimportant as far as hospital functions were concerned. From the number of entries in the telephone directories the height of popularity of nursing homes seemed to be in the 1930s, but thereafter there was a notable decline in their number.

C2 Dates of hospital origin

Terms such as 'built', 'instituted', 'founded', 'opened', or similar conventions entered against hospitals listed in different sources were taken to have a similar meaning in fixing the dates of origin of different hospitals. This is inaccurate as inevitable time-lags exist between the conception and commissioning of a hospital, but these terms were usually the only indications of age available. Burdett provided a good coverage in the sections of the directories labelled 'London hospitals' and 'provincial hospitals'; for other hospitals, the date of first printed appearance in a volume was assumed. This left two other major categories of hospitals partly unaccounted for: hospitals for infectious diseases, and hospitals opened in the 1930s and 1940s. Dates of origin of the first group were obtained from *Return on sanitary districts: accommodation for infectious diseases* (House of Commons 1895); dates for the second category were obtained, where possible through personal contact with the hospitals concerned.

C3   Classification of hospitals

Type 1 hospitals (acute) were generally defined on the basis of the proportion of beds allocated to general medical and other acute services based on the Department of Health and Social Security's classification given in *Notes on hospital form SH3 for 1976* (DHSS 1976). This would be equivalent to the categories of 'acute', 'mainly acute' and 'partly acute', but it would also include maternity hospitals. The word 'proportion' was problematic as figures were not always provided in older sources to establish the size of the proportion. In such cases, the section of the directory, the name of the hospital, a short description of its work, and other corroborative material was usually sufficient to make the correct identification.

Type 2 hospitals provide specialised services. They are defined as those hospitals engaged in acute work for which the largest proportion of bed accommodation is devoted to medical services related to one of the following specialties:

(a)   diseases of the throat, nose, skin or ear
(b)   diseases of the heart and chest
(c)   diseases of the eyes
(d)   urological disorders
(e)   children's ailments
(f)   gynaecological disorders (but not general women's hospitals)
(g)   acute nervous diseases
(h)   non-convalescent orthopaedic disorders
(i)   miscellaneous disorders such as tropical and venereal diseases.

Excluded from the specialist category were hospitals with religious, ethnic, or other special affiliations unless they specialised in any of the above.

Decisions on which hospitals to include in the type 3 category at different times were problematic due to changes in the care, treatment and definition of mental diseases. For practical reasons they were usually based on information provided by the directories as these were considered to be most in touch with the hospital system at the time of their publication. According to a publication by the DHSS (e.g. DHSS 1976), a psychiatric hospital is one in which 90 per cent or more of available beds are used for patients with mental illness. The definition includes child and adolescent psychiatry, mental hospitals for the criminally inclined, as well as conventional psychiatric hospitals and hospitals for the mentally handicapped. Not included, however, are certain homes, special schools and related establishments, even though these are sometimes listed in the hospital sections of old directories, under mental institutions.

Mental illness has created different images in the minds of the public and medical profession, and indicative of the change in attitudes are the terms describing the type of institution dealing with the mentally ill. Last century, for example, mental hospitals were variously described as asylums, lunatic asylums and institutions for mental defectives or the insane; nowadays they are psychiatric hospitals. These changes also reflect, of course, large differences in the diagnosis and subsequent treatment of the mentally ill, so that today it is easier to draw a distinction between hospitals, special homes and so forth. For all of these reasons, it was hence decided to adhere to contemporary judgement and descriptions, rather than to impose in retrospect the rather stricter definitions in use today.

Type 4 hospitals are long stay institutions other than those dealing with mental illness. In this category are hospitals with the largest proportion of their beds devoted to one or more of the following clinical specialties: chronic diseases of the chest, geriatry, rehabilitation and miscellaneous chronic disorders. Thus included in this category are sanatoria for the treatment of tuberculosis (but not 'isolation hospitals'), homes and hospices for incurables, convalescent hospitals, hospitals for the aged and infirm, long-stay hospitals for orthopaedic medicine and retreats for alcoholics and lepers.

Type 5 hospitals are those hospitals specialising in the treatment of infectious diseases. The authority originally responsible for this group of hospitals was again the Metropolitan Asylums Board (MAB, 1867–1930). Before the inception of the MAB few facilities existed, many hospitals refusing to allocate accommodation to sufferers of infectious diseases (Abel-Smith 1964). Establishments outside the jurisdiction of the MAB were administered by local Poor Law unions. The list of infectious diseases encountered in these hospitals was long; the most significant were cholera, smallpox, enteric fever, measles, typhus, diphtheria, whooping cough and scarlet fever, although at any time some tended to be much more prominent than others. Early this century there were numerous hospitals treating infectious diseases, but their numbers have since dwindled because of advances in the treatment and isolation of cases and because of improved public health regulations. The most famous hospitals were those located on the lower Thames near Dartford, south-east London, and the four Metropolitan Fever Hospitals located nearer the centre of London, which today are devoted to other uses.

C4   Administrative categories of hospital

Two classes of administrative category, private and public, were distinguished depending on the mode of finance of the hospital concerned. Before 1948 hospitals were classed as 'independent' if they received financial support from endowments, charities, philanthropists, insurance companies and others. Although such hospitals received an increasing proportion of their revenue from government sources prior to 1948, there existed a clear distinction between them and hospitals in the 'public' sector. The latter hospitals were mostly of types 3, 4 and 5, whereas independent hosptals tended to be of either types 1 or 2. After 1948, hospitals were invited to join the newly created NHS, with the majority of hospitals in the 'independent' sector responding. For the purposes of the case study, therefore, hospitals in the formerly 'public' sector were termed 'state-run' hospitals, while those left in the independent sector were termed 'private' hospitals.

C5 Hospital locations

Whereas dates of origin were recorded on a continuous basis, the locations and sizes of hospitals were obtained at five points in time – 1901, 1921, 1939, 1951 and 1971, each year corresponding to the population census (or national register in the case of 1939). Hospitals were located on a map using a six-figure Ordnance Survey map reference. To avoid the possibility of double counting, or of making other errors, hospitals were usually recorded by their original names.

Whereas the majority of hospitals consist of clusters of buildings, the map coordinates of a hospital are for one point in space. To identify the locations of individual hospitals, specific procedures had therefore to be adopted. Firstly, if a hospital had a separate name or identity it counted as one hospital, regardless of whether another hospital was located in the immediate vicinity. Secondly, an annex to a hospital was counted as a separate hospital if it was located more than a reasonable distance from the parent hospital and if it was clear that the annex performed an important role. In some cases, difficulties arose where source addresses in directions were insufficient to locate a hospital on a map. Here, either alternative information was sought or reasonable assumptions were made based on the best information available.

C6 Hospital size

A variety of measures of hospital size have been used in hospital studies, but none is entirely satisfactory. The main advantage of using the number of available beds to measure size is that it is the most frequently stated statistic in historical data sources. Also it is highly correlated with alternative measures based on the level of hospital activity (e.g. case-loads, inpatient-days, surgical operations). A disadvantage is that the density of beds in hospital wards varies from hospital to hospital and, particularly in historical instances, high densities are usually construed as an indication of lower standards and hence quality of treatment. Another disadvantage is that the number of beds provides little indication of the sophistication of the medical techniques employed (although, as a general rule, the larger hospitals tend to be better equipped). Finally, again from the historical viewpoint, some hospitals may have inflated their numbers of beds in order to attract more charitable finance. Plainly, these considerations affect to some degree the analysis of trends and variability in hospital size and the consequent conclusions given in chapter 5. These have been taken into account wherever possible.

C7   Summary of hospital statistics

Tables C1–C8 give the summary statistics for each category of hospital located in the 50 km radius defined in the case study. They show the number of hospitals at each date, the mean, median and modal sizes, the standard deviation, skewness, and kurtosis and the size-range of the facilities.

Table C.1   All hospital types.

	1901	1921	1939	1951	1971
hospital total	339	455	549	546	509
mean size (beds)	190.09	230.61	270.34	278.86	254.07
median	40.0	50.0	86.0	100.0	109.0
mode	0–50	0–50	0–50	0–50	0–50
standard deviation	393.07	423.03	456.68	463.13	375.54
skewness	3.97	3.18	3.09	3.08	2.92
kurtosis	18.63	11.21	10.37	9.99	9.27
minimum size	4	4	6	6	6
maximum size	2876	2605	2747	2702	2227

Table C.2   Type 1 hospitals.

	1901	1921	1939	1951	1971
hospital total	152	196	260	277	257
mean size	155.71	179.44	206.46	215.49	214.55
median	33.5	48.5	74.0	85.0	128.0
mode	0–50	0–50	0–50	0–50	0–50
standard deviation	226.13	260.53	280.77	267.14	1.39
skewness	1.50	2.01	1.96	1.98	1.39
kurtosis	0.78	4.57	3.53	4.29	1.36
minimum size	5	6	7	7	12
maximum size	776	1600	1420	1514	1040

Table C.3   Type 2 hospitals.

	1901	1921	1939	1951	1971
hospital total	43	54	57	47	47
mean size	59.53	111.07	125.14	137.85	123.57
median	45.0	54.5	80.0	85.0	85.0
mode	0–50	0–50	51–100	51–100	51–100
standard deviation	62.35	153.73	177.96	200.86	135.34
skewness	2.38	3.14	5.04	4.28	2.68
kurtosis	5.97	10.37	29.35	20.62	8.09
minimum size	6	6	10	6	6
maximum size	321	810	1284	1284	736

Table C.4   Type 3 hospitals.

	1901	1921	1939	1951	1971
hospital total	33	41	64	71	72
mean size	809.67	1033.78	927.97	914.30	725.03
median	400.0	810.0	625.0	510.0	369.5
mode	1000	1000	1000	1000	1000
standard deviation	902.53	858.92	921.49	909.16	723.92
skewness	0.95	0.36	0.60	0.61	0.63
kurtosis	−0.56	−1.34	−1.24	−1.26	−1.13
minimum size	12	28	10	14	12
maximum size	2876	2605	2747	2702	2227

Table C.5   Type 4 hospitals.

	1901	1921	1939	1951	1971
hospital total	51	69	100	119	130
mean size	78.51	119.55	155.77	130.76	123.58
median	30	50.0	70.5	75.0	82.5
mode	0–50	0–50	0–50	0–50	0–50
standard deviation	128.70	150.65	184.51	147.29	116.73
skewness	2.71	1.65	1.84	1.84	1.71
kurtosis	6.61	1.64	3.51	2.70	2.55
minimum size	5	10	9	10	10
maximum size	600	600	981	700	564

Table C.6   Type 5 hospitals.

	1901	1921	1939	1951	1971
hospital total	60	95	68	32	3
mean size	124.85	138.17	185.87	175.5	35.33
median	29.0	40.0	96.5	86.5	40.0
mode	0–50	0–50	5–100	0–50	0–50
standard deviation	224.43	259.02	228.42	201.44	116.73
skewness	2.65	3.15	1.93	1.65	1.71
kurtosis	7.71	10.75	3.26	1.44	2.55
minimum size	4	4	6	20	10
maximum size	1192	1524	1104	756	564

Table C.7   Independent/private hospitals.

	1901	1921	1939	1951	1971
hospital total	225	258	300	54	48
mean size	70.44	74.84	100.06	86.48	82.73
median	30.0	36.0	58.0	52.0	55.0
mode	0–50	0–50	0–50	0–50	0–50
standard deviation	115.69	109.92	124.62	91.02	68.60
skewness	3.57	3.51	3.11	2.48	1.54
kurtosis	14.86	14.50	12.02	6.94	2.15
minimum size	5	6	9	10	10
maximum size	776	779	891	489	316

Table C.8   Public/state hospitals

	1901	1921	1939	1951	1971
hospital total	114	197	249	492	461
mean size	426.25	434.62	475.51	299.98	271.91
median	272.0	204.0	210.0	105.0	120.0
mode	0–50	0–50	0–50	0–50	0–50
standard deviation	592.31	570.00	603.98	482.36	389.73
skewness	2.22	1.93	1.93	2.91	2.76
kurtosis	5.00	3.42	3.30	8.73	8.16
minimum size	4	4	6	6	6
maximum size	2876	2605	2747	2702	2227

References

Abel-Smith, B. 1964. *The hospitals 1800–1948*. London: Heinemann.

Alonso, W. 1965. *Location and land use*. Cambridge, Mass.: Harvard University Press.

Angel, S. and G. Hyman 1976. *Urban fields*. London: Pion.

Arrow, K. J. 1963. Uncertainty and the welfare economics of medical care. *American Economic Review* **53**, 941–73.

Ayers, G. M. 1971. *Englands first state hospitals, 1867–1930*. London: Wellcome Institute of the History of Medicine.

Beavon, K. S. O. 1977. *Central place theory: a reinterpretation*. London: Longman.

Burdett, H. 1889–1930. *Burdett's hospitals and charities: the year book of philanthropy and hospital annual*. London: The Scientific Press. (Continued after 1930 as *Hospitals yearbook*.)

Bussiere, R. 1972. Static and dynamic characteristics of the negative exponential model of urban population distributions. In *Patterns and processes in urban and regional systems*. A. G. Wilson (ed.). London: Pion.

Bussiere, R. and T. Stovall 1981. *Systèmes evolutifs urbains et regionaux a l'état d'equilibre*. Paris: SPM.

Christaller, W. 1960. *Central places in southern Germany* (translated by C. W. Baskin). Englewood Cliffs. NJ: Prentice-Hall. (Originally published as *Die zentralen Orte in Suddeutschland*. Jena: Fischer, 1933.)

Clark, C. 1951. Urban population densities. *Journal of the Royal Statistical Society, Series A* **114**, 490–6.

Clayton, K. M. (ed.) 1964. *Guide to London excursions*. 20th International Geographical Congress.

Cohen, H. A. 1967. Variations in cost among hospitals of different sizes. *Southern Economic Journal* **33**, 355–66.

Culyer, A. J. 1971. The nature of the commodity 'health care' and its efficient allocation. *Oxford Economic Papers* **23**, 189–211.

Dainton, C. 1961. *The story of England's hospitals*. London: London Museum.

DHSS 1976. *Notes on hospital form SH3 for 1976*. London: DHSS.

Dietrich, C. 1977. *Macromodels of inpatient and outpatient health care systems in the Federal Republic of Germany*. Laxenburg, Austria: International Institute for Applied Systems Analysis.

Eckstein, H. H. 1958. *English health service: its origins, structure and achievements*. Cambridge, Mass.: Harvard University Press.

England and Wales Board of Control 1938. *A list of state institutions, certified houses and approved homes for mental defectives*. London: HMSO.

Feldstein, M. S. 1963. Economic analysis, operational research, and the National Health Service. *Oxford Economic Papers* **15**, 19–31.

Feldstein, M. S. 1965b. Hospital cost variation and case-mix differences. *Medical Care* **3**, 95–103.

Fletcher, R. 1971. A general quadratic programming algorithm. *Journal of the Institute of Mathematics and its Applications* **7**, 76–91.

Fuchs, V. R. 1966. The contribution of health services to the American economy. *Millbank Memorial Fund Quarterly* **44**, 65–101.

Gross, P. F. 1972. Urban health disorders, spatial analyses and the economics of health facility location. *International Journal of Health Services* **2**, 63–84.

Hall, N. 1982. *Modelling patient flows in the Lower Hunter*. MSc thesis, Faculty of Mathematics, University of Newcastle, NSW, Australia.

HMSO 1944. *A National Health Service*. Cmd 6502. London: HMSO.

HMSO 1945. *Hospital survey*. London: Ministry of Health.

HMSO 1978. *Hospital inpatient enquiry for England and Wales*. Series MB4, Number 12. London: HMSO.

House of Commons 1895. Return on sanitary districts: accommodation for infectious diseases. *Hansard*, House of Commons Sessional Paper 28, vol. LXXX. London: HMSO.

Hyman, G. and L. Mayhew 1982. *On the geometry of emergency service medical provision in cities*. Laxenburg, Austria: International Institute for Applied Systems Analysis.

Hyman, G. and L. Mayhew 1983. On the geometry of emergency service medical provision in cities. *Environment and Planning A* **15**, 1669–90.

Illich, I. 1976. *Limits to medicine*. Harmondsworth: Penguin.

Lewis, P. W. 1977. *Maps and statistics*. London: Methuen.

LCC 1901, 1921. *Records of the Boards of Guardians, form A weekly returns*. Passed to LCC (now GLC), County Hall, London.

LCC 1929. *Survey of accommodation at schools, cottages and receiving homes*, vol. IV–V. Local Government Act 1929. London: Architect's Department, LCC.

LCC 1929. *General survey of children's institutions and mental hospitals*. Metropolitan Asylums Board.

LCC 1929. *Casual wards and miscellaneous buildings*. Local Government Act 1929. London: Architect's Department, LCC.

LCC 1929. *Survey of hospital and institutional accommodation*, vol. I–III. Local Government Act 1929. London: Architect's Department, LCC.

LCC 1933. Part II of a joint survey of medical and surgical services in the administrative county of London – municipal hospitals, clinics and dispensaries. London: LCC.

LCC 1949. *The LCC hospitals: a retrospect*. London: LCC.

LHPC 1979. *Acute hospital services in London: a profile by the London Health Planning Consortium*. London: HMSO.

Mayhew, L. 1979. *The theory and practice of urban hospital location*. PhD thesis, Birkbeck College, University of London.

Mayhew, L. 1980. *The regional planning of health care services: RAMOS and RAMOS⁻¹*. Laxenburg, Austria: International Institute for Applied Systems Analysis.

Mayhew, L. 1981. Automated isochrones and the locations of emergency medical services in cities. *Professional Geographer* **33**, 423–8.

Mayhew, L. and G. Leonardi 1982. Equity efficiency and accessibility in urban and regional health care systems. *Environment and Planning A* **14**, 1479–507.

Mayhew, L. and G. Leonardi 1984. Resource allocation in multi-level spatial health care systems. In *Papers and Proceedings of the Regional Science Association*. London: Pion.

Mayhew, L. and A. Taket 1980. *RAMOS: a model of health care resource allocation in space*. Laxenburg, Austria: International Institute for Applied Systems Analysis.

Mayhew, L. and A. Taket 1981. *RAMOS: a model validation and sensitivity analysis*. Laxenburg, Austria: International Institute for Applied Systems Analysis.

Mayhew, L. and T. Bowen 1984. *The potential for day hospitals in Piemonte: a feasibility study*. Turin, Italy: IRES.

Mayhew, L., R. W. Gibberd and H. Hall 1985. Predicting patient flows and hospital case-mix. Forthcoming *Environment and Planning A*.

Mills, E. S. 1970. Urban density functions. *Urban Studies* 7, 5–20.

Muth, R. F. 1961. The spatial structure of the housing market. In *Papers and Proceedings of the Regional Science Association*. London: Pion.

Newhouse, J. P. 1970. Towards a theory of non-profit institutions: an economic model of a hospital. *American Economic Review* 60, 64–74.

Park, R. E. and E. W. Burgess 1925. *The city*. Chicago: University of Chicago Press.

Parr, J. B. 1980. Health care facility planning: some developmental considerations. *Socio-economic Planning Sciences* 14, 121–7.

Pinker, R. 1966. *English hospital statistics 1861–1938*. London: Heinemann.

Powell, A. 1930. *The Metropolitan Asylums Board and its work, 1867–1930*. London: The Metropolitan Asylums Board.

Reder, M. W. 1965. Some problems in the economics of hospitals. *Economic Review* 55, 472–80.

RAWP 1976. *Sharing resources for health in England*. Report of the Resource Allocation Working Party. London: HMSO.

Rising, E. and L. Mayhew 1983. *The spatial allocation of medical care resources in Massachusetts: an application of RAMOS*. Laxenburg, Austria: Institute for Applied Systems Analysis.

Schultz, G. P. 1970. The logic of health care facility planning. *Socio-economic Planning Science* 4, 383–93.

Shannon, G. W. and A. G. E. Denver 1974. *Health care delivery: spatial perspective*. New York: McGraw-Hill.

Singer, C. and E. A. Underwood 1962. *A short history of medicine*, 2nd edn. Oxford: Clarendon Press.

Tadei, R. *et al.* 1983. *Un'amalisi con il modello RAMOS della struttura spaziale del servizio sanitario regionale: il caso del Piemonte*. Turin, Italy: IRES.

Tobler, W. R. 1963. Geographic area and map projections. *Geograhical Review* 53, 59–78. Reprinted in Angel and Hyman 1976, *q.v.*

United Nations 1980. *Patterns of urban and rural population growth*. New York: United Nations.

Weisbrod, B. A. 1968. *Economics of public health: measuring the economic impact of diseases*. University Park, PA: University of Pennsylvania Press.

Wilson, A. G. 1974. *Urban and regional models in geography and planning*. London: Wiley.

Wingo, L. Jr 1961. *Transportation in urban land: resources for the future*. Baltimore: John Hopkins University Press.

World Health Organisation 1963. *A framework for international health accounting*. Geneva: WHO.

Zeilinski, K. 1980. The modelling of urban population density: survey. *Environment and Planning A* 12, 135–54.

Index